DESERT VEGETATION
OF ISRAEL AND SINAI

AVINOAM DANIN

DESERT VEGETATION OF ISRAEL AND SINAI

Avinoam Danin Ph.D.
Department of Botany
The Hebrew University, Jerusalem

Cana Publishing House

The maps in this book were prepared from material of the Survey of Israel and by their permission.
Figure 8 is presented by the courtesy of NASA.
Maps and diagrams were drawn by Mrs. Tamar Soffer.
Photographs — Figures No. 18, 20, 21, 22, 34 and 122 are by Mr. Jacob Gamburg; Figure 27 by Peter Grossman; Figures 48 and 100 by Mrs. Esther Huber; The rest by the author.

Partialy based on Danin, A., 1977, The Vegetation of the Negev (north of Nahal Paran), Sifriat Poalim & Yahdav in Hebrew).

Editing: Dr. Jeanette Greenberg
Dr. Philip Alkon
Design and cover: Karp Studio, Jerusalem
Phototypesetting: Yeda-Sela, Tel-Aviv
Phototypesetting: Efrat, Jerusalem
Color seperation: Spectra, Jerusalem
Films and plates: Printon, Jerusalem
Printing: Golan Press, Jerusalem
Binding: Kerech, Jerusalem

ISBN 965-264-005-0
Printed in Jerusalem, Israel

CONTENTS

PREFACE

The ability of plants to survive and even flourish under the harsh conditions of desert environments, has long fascinated botanists and amateur naturalists. What adaptations enable plants to exist and thrive in arid zones? How do individual plants and populations compete for scarce water and suitable habitat? How do biotic and abiotic features of the environment shape the distribution and structure of desert plant communities? Do the survival mechanisms of desert plants differ from those of other ecosystems? These and similar questions have captured the attention of students of desert life throughout the world.

The potential development of the earth's vast desert areas for agriculture and other human needs also demands an awareness of the ecological characteristics and requirements of desert vegetation.

Material in this volume pertaining to the Negev and Judean Deserts is derived from previously published studies (18, 30, 31, 84 in the reference list). Parts of the present book are translated from the Hebrew edition (25). The information on Sinai vegetation is based on the work of a research team of the Department of Botany, The Hebrew University of Jerusalem, directed by Professor G. Orshan. In addition to the author, participants included Professor N.H. Tadmor, Dr. G. Halevy and Dr. A. Shmida.

Amygdalus ramonensis an endemic tree of the Central Negev Highlands.

Ferula daninii named by M. Zohary after A. Danin who discovered this plant.

INTRODUCTION

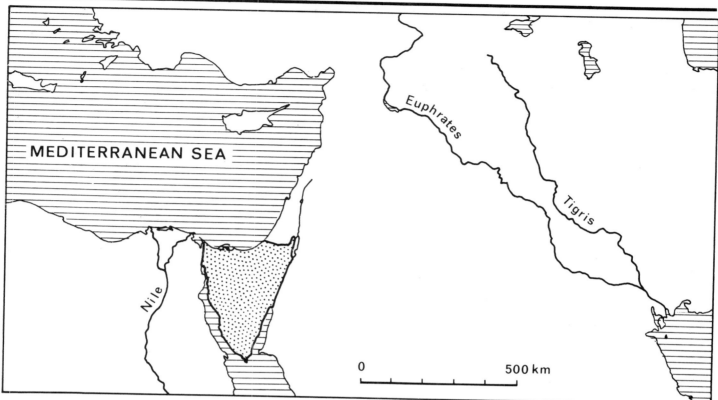

MEDITERRANEAN SEA

Euphrates

Tigris

Nile

0   500 km

The deserts of Israel and Sinai belong to the Arabian type (91). They are characterized by:
1. Arid to extremely arid climate with a Mediterranean influence
2. Precipitation mostly in winter
3. Mean temperature of 10° to 20°C in the coldest month and 20° to 30°C in the warmest month.

The deserts we studied (Fig. 1) form the northern boundary of the subtropical belt of deserts. The desert of Sinai becomes narrow in the Negev and even narrower in the Judean Desert.

Four plant geographical regions meet in this part of the Middle East (144). These are:
1. Mediterranean, with forest and shrub vegetation
2. Irano-Turanian, with steppe vegetation
3. Saharo-Arabian, with desert vegetation

Fig. 1. Map showing location of the area under study.

4. Sudanian penetration with savanna-like vegetation.

These four regions superimposed on the great diversity of climate, rock and soil types make possible the existence of some 1,300 species in this area of 74,300 km². This figure is comparable to the 1,666 species recorded from the British Isles which cover an area of 229,850 km². (68).

In this book, the distribution of plant communities in the Israeli and Sinai deserts is discussed in relation to environmental conditions. Some adaptations of species to these deserts are presented. Listed here are plants important for man in the desert; they may be edible, poisonous or serve as indicators for water. An illustrated key to the 200 most common trees, shrubs and semishrubs is presented as well.

CHAPTER 1.
ENVIRONMENTAL CONDITIONS

CLIMATE

Climatic data are more complete for the Negev than Sinai. General seasonal patterns of temperature, precipitation and relative humidity are similar to those in the Mediterranean region. The summers are hot and dry and the winters are cool. Rain falls almost exclusively in winter. However, precipitation and relative humidity are lower and temperatures higher in the desert than in the adjacent Mediterranean region. Summer rainstorms, south of the Suez-Elat line, result from barometric depressions coming from the Red Sea. Although rare, these rains have an important impact on plant life in the area. Prevailing temperatures are lower and precipitation greater at high elevations in southern Sinai.

Seasonal and daily fluctuations in climatic conditions are much greater in deserts than in more temperate regions. Thus, environmental conditions are extremely variable as well as harsh in the arid zones.

PRECIPITATION

Mean annual rainfall in these deserts ranges from 250 mm at the northern boundary of the Negev to only 10—20 mm in the southern Sinai. In the Negev, annual rainfall decreases from north to south (Fig. 2). The Judean Desert, which lies in the rain shadow of the Judean Mountains, is characterized by a steeply declining rainfall going towards the Dead Sea. This gradient can be demonstrated by the density of isohyets in Fig. 2.

Most of the southern Negev and Sinai receive less than 50 mm of rainfall annually. However, in mountainous areas the rising air masses become cooler and thus relative humidity increases. Therefore, mountain ranges such as Gebel Halal, Gebel el 'Igma and the southern Sinai Massif receive more precipitation (48,73).

Precipitation may occur as snow on the high peaks of southern Sinai, and winter snow lasting two to four weeks has been observed on the northern slopes of Gebel Katherina. In some years more than one snowfall may occur while in other years snow may be absent. Precipitation which falls as rain in the valleys of southern Sinai may occur as hail on the high peaks. Water derived from melting snow or hail is more likely to infiltrate the desert soil because of its low rate of percolation.

In dry years the isohyets shift to the north and west, while in wet years they shift to the south and east. In low precipitation areas, where rainfall varies considerably from year to year, the rainfall map is least reliable. Mean data may also be misleading in characterizing precipitation because of the nature of desert rainstorms. Considerable precipitation occurs as a result of convective rains (118), which are very local in extent and irregular in occurrence. For example, a large area along the Gulf of Elat near Sharm el Sheikh

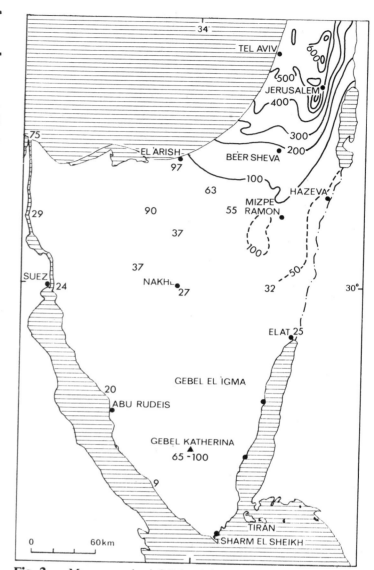

Fig. 2. Mean annual rainfall (mm) (based on 3 and 48).

received effective rains in 1968, resulting in the widespread growth of the scented composite *Pulicaria desertorum*. In 1970, by contrast, most of the *P. desertorum* in the area were dead due to the lack of effective precipitation. In 1971 effective rains fell again, but this time only in small areas (cf. Sharon, 117). The boundaries of the rain could be mapped by the occurrence of living *P. desertorum* plants. Similar green "islands" in the desert are a common sight throughout much of Sinai.

The number of convective rains per season is unpredictable. In parts of the 'Arava Valley and Sinai floods resulting from convective rains have been observed during all seasons. Summer rains resulting from the influence of the Red Sea depressions (48) cause floods. Tropical plants of Sudanian origin may germinate in the moist wadi beds following summer rains. Although such rains are rare, the resultant germination of tropical shrubs and trees marks a long-term change in the nature of the vegetation.

Analyses of tree ring data (dendrochronology) suggest climatic cycles having a periodicity of about 100 years in the

SAND

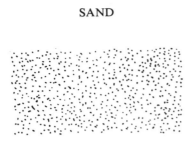

All of the rain water infiltrates the soil. The amount of water available to the plants is greater than in fine-grained soils.

CHALK and LOESS

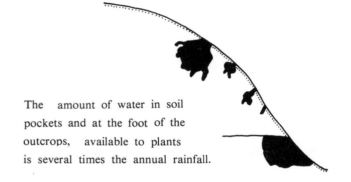

There is a loss of water by runoff and direct evaporation. 30-50 percent of the rainfall is available to the plants. In wadis the amount of available water is several times the annual rainfall.

HARD-BEDDED LIMESTONE, FISSURED GRANITE AND METAMORPHIC ROCKS

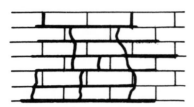

60-90 percent of the rainfall is available to the plants, water concentrates in fissures of the hard rocks and in soft interbeds.

SMOOTH-FACED OUTCROPS OF MASSIVE ROCKS

The amount of water in soil pockets and at the foot of the outcrops, available to plants is several times the annual rainfall.

Fig. 3. The influence of rock and soil on water regime under desert conditions.

northern Negev (116). As compared with the long-term mean annual rainfall (84 mm), years of maximum rainfall averaged 110 mm and years of minimum rainfall 58 mm. In the northern Sinai, where the long-term mean annual rainfall is 100 mm, a climatic cycle with a periodicity of 300 mm may exist. Here the years of maximum rainfall averaged 300 mm and years of minimum rainfall 30 mm (126).

Fog and dew are essential to many desert plants, especially during the summer and on dry winter days. In the Central Negev Highlands annual dew accumulation ranges from 26 to 36 mm with dew occurring on approximately 200 nights each year. In some years dew provides more moisture than does rainfall. Lichens are particularly effective in capturing fog and dew. They carry on photosynthesis during morning hours using moisture absorbed at night (70). Summer fog and dew are frequent in Sinai north of Gebel et Tih, but absent in the southern Sinai. At Gebel et Tih, fog and dew sometimes thoroughly wet the soil and rocks. Along the Suez Canal, the soil was observed to be wet to depths of 0.5 to 1.0 cm in summer as a result of fog and dew.

Vascular plants in the Namibian Desert use soil moisture which is supplied by fog (129). In parts of the South American

coastal desert, vegetation depends solely on fog as a source of moisture.

TEMPERATURE

Mean annual temperatures in these deserts range from 9°C to 25°C. The lowest temperatures prevail in Gebel Katherina in Sinai (2,642m above sea level). The highest temperatures are found in the Dead Sea Valley, the 'Arava Valley and along the Gulf of Elat, elevation 400 m below sea level to 200 m above sea level (3,48).

At the Dead Sea Valley and 'Arava Valley monthly mean temperatures range from a low of 13° to 16°C in January to a high of 31° to 34°C in August. At Gebel Katherina mean monthly temperatures range from (-)1° to 2°C in winter to 17° to 19°C in summer. The coldest area in the Negev is Har Ramon (elevation 1,035 m). The mean annual temperature is 17°C, ranging from a mean of 8°C in January to 24°C in August. In general, the lower the elevation the higher the mean temperatures. Further information on temperatures will be found in the description of each "vegetation district" (chap. 3).

Whereas plant species distribution have been found to be affected by mean temperatures, no such correlation has yet been found with temperature extremes.

Plate 1.   Sands in southern Sinai. On the left a thick sand cover enables the development of semishrubs throughout this area; on the right, silt in the alluvium derived from the weathering of the black basalt causes a water regime which enables plants to grow only in the wadis.

Variations in slope direction and soil type result in microhabitats which differ substantially in temperature regime. South-facing slopes are subject to prolonged exposure to solar radiation and thus have a relatively warm microclimate. Dark surfaces such as magmatic, metamorphic and chert rocks absorb solar radiation more efficiently and have warmer microclimates than lighter-colored rocks.

EDAPHIC CONDITIONS (SOIL AND ROCK)

Soil is defined here as the weathered upper part of the earth's crust subject to penetration by plant roots. This definition encompasses rocky areas that serve as a reservoir of water and essential minerals utilized by plants.

Several desert soil types can be recognized according to their water absorbing capacity and salt regime (Fig. 3). The following soil classification is based on the parent rock from which the soil is derived and emphasizes characteristics of

importance to plant life. Figure 4 presents the main locations of each edaphic type. Because of the complicated pattern of rock outcrops, a rough generalization was made.

SANDS

The rough texture of sand causes rainwater to infiltrate easily. However the water holding capacity of sand is low. Much of the water may percolate to depths beyond the root zone and thus become unavailable to desert plants. There is less infiltration of rainwater in sands containing silt and clay such as occur at the edge of the sands in the Negev foothills and near basalt outcrops in southern Sinai. In these areas precipitation may be lost as silt-laden runoff (Plate 1). In certain places in southern Sinai and the 'Arava Valley, *Acacia raddiana* trees are restricted to those wadis having a mixture of silt, clay and sand. Sand poor in silt and clay and therefore having a lower water holding capacity do not support such *Acacia* trees (Plate 11).

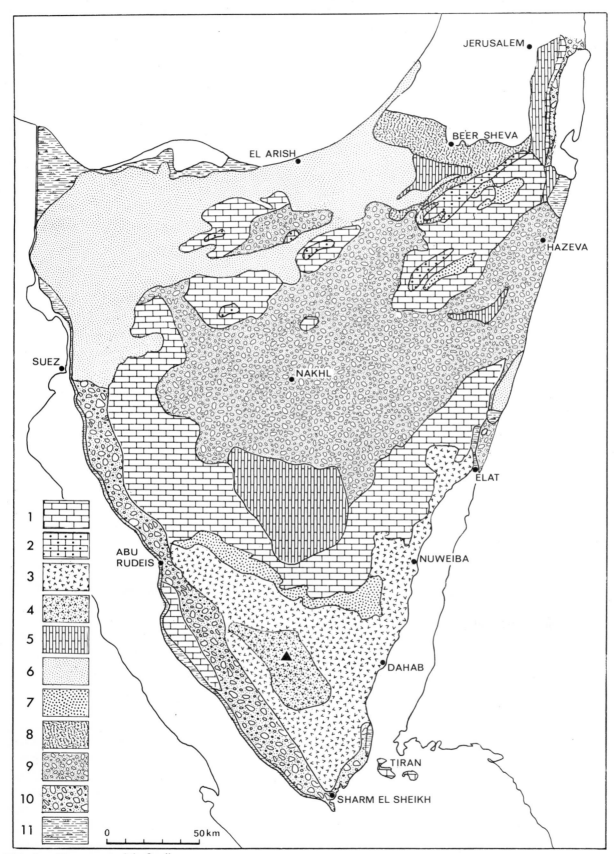

Fig. 4. Parent material of soils.

1. Limestone and dolomite (in Tiran with gypsum).
2. Limestone with large outcrops of smooth-faced rocks.
3. Magmatic and metamorphic rocks.
4. Magmatic rocks with outcrops of smooth-faced granite.
5. Chalk and soft sedimentary rocks.
6. Sand.
7. Old sandstone with or without sand cover.
8. Loess.
9. Gravel plains and alluvium.
10. Alluvium.　11. Salt marshes.

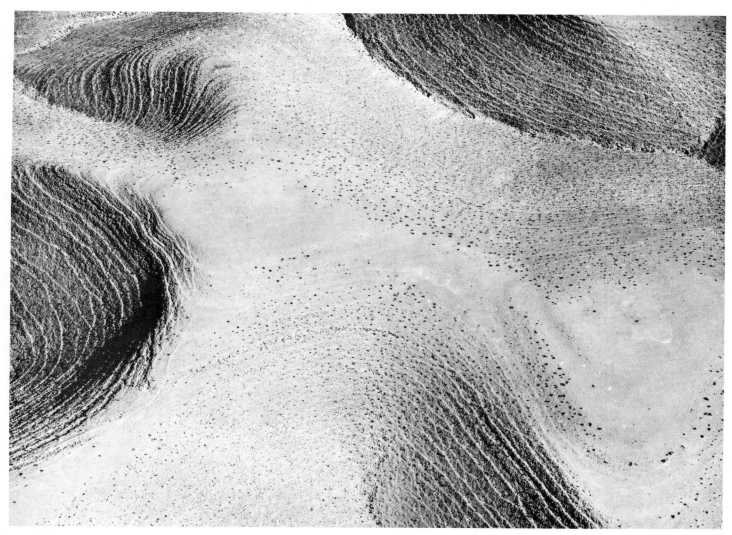

Fig. 5. Aerial photograph showing slope vegetation of chalk hill in Gebel el 'Igma. The white lines on the vegetated parts of the slope are paths made by grazing goats and sheep.

LOESS AND REGS (GRAVEL PLAINS)

Loess is a deposit of windblown silt and clay with some sand. It occurs mainly in the valleys and plateaux of the Negev and northern Sinai (137). In central Sinai and the southern Negev there are large expanses of gravel plains where silty soil is covered by chert gravels. This soil type is known as reg (15) or the hammadas (110). An important characteristic of these substrates is the formation of a surface crust following rainfall due to the reorganization of soil minerals. In many cases a biological crust of algae and lichens is also formed, which seals the soil by absorbing water and swelling during rain. In such cases, 70 percent of the rainfall may be lost by runoff and evaporation. Water which penetrates the surface of loess soil is stored near the surface due to the high water holding capacity of this fine-grained soil. Subsequent rains usually do not arrive in time to bring the capillary water to a depth below 20 cm, so that much of the soil water is lost by evaporation. Rain water contains small quantities of salts which concentrate in the soil because of

evaporation (136). The depth of the salt layers is influenced by soil texture and the amount of rainfall at the particular site. Gypsum, calcium carbonate and sodium chloride are the principal salts involved. Because of the high rate of runoff and evaporation only about 30 to 50 percent of the rain that falls on these soils is available to the plants following evaporation and runoff.

The different kinds of loess soil can be distinguished by the proportion of sand, gravel and pebbles mixed with the fine-grained silt and clay, and by their salinity.

CHALK, MARL AND CLAY

These sedimentary rocks are composed of fine-grained components and are characterized by a loess-like surface material which accumulates as the result of weathering. A biological crust often occurs on these soils and results in high runoff and evaporation as described above. A clay-rich marl which occurs in the central Negev badlands near Sede Boqer and at the Gebel el 'Igma escarpment in Sinai is weathered to a depth of 5-10 cm; this is also the mean depth of water penetration. Most precipitation is absorbed by this shallow layer and is quickly evaporated, resulting in salt accumulation as with loess soils. Thus, this soil supports only a sparse vegetation (Fig. 5).

Clay outcrops are the poorest substrate for vegetation in the desert. The amount of water available to plants in various kinds of chalk is higher than in marl and clay, but even in chalk a large proportion of water is lost through evaporation and runoff. It is estimated that only 20 to 50 percent of the precipitation is available to plants in soils derived from chalk, marl or clay, depending upon rock type. Most of these soils lack the deep fissures which facilitate water penetration and provide protection from evaporation. In this respect they resemble loess soils.

BEDDED LIMESTONE, DOLOMITE, FISSURED MAGMATIC AND METAMORPHIC ROCKS

These substrates are formed from weathered outcrops of densely fissured hard rocks. The fissures in this substrate act as paths through which water can freely penetrate to a considerable depth. Also, the surface rocks and stones form discontinuities which prevent the extensive physical and biological crusting typical of fine-grained soils. Most rainfall passes into the fissures and into the soft layers between the hard strata. This moisture is protected from direct evaporation by the covering rocks.

Geomorphological studies indicate that coarse-grained surface material provides for maximum soil water penetration and minimum runoff (139). From 60 to 90 percent of the rainfall is available to plants in these soil types.

CLIFFS AND SMOOTH-FACED ROCK OUTCROPS

Large outcrops of smooth-faced rocks are associated with limestone, dolomite, granite and sandstone deposits. This substrate exhibits a water regime that contrasts markedly with other soil types. Hard smooth-faced rocks absorb less than 2 percent of their weight in water and lack small depressions that can store water (31). Thus, even weak showers result in runoff. Flow water first reaches the few existing crevices, fissures and soil pockets in the rock outcrop. With sufficient rainfall, the excess water next flows to the soil at the base of the outcrops. Only after this habitat is saturated, does runoff continue to the wadis. The amount of water that accumulates in the soil pockets and at the bases of the outcrops amounts to several times the annual rainfall. Consequently, the number of rains that can be used by the plants is much higher than in other soil types exposed to the same gross precipitation. Showers which in loess soil infiltrate to a depth of a few centimeters and evaporate shortly thereafter cause runoff on the smooth rock faces and saturate the available soil in crevices and soil pockets (Fig. 3). These areas, thus, comprise relatively humid habitats capable of supporting Mediterranean species such as *Narcissus tazetta*, *Sternbergia clusiana* and *Juniperus phoenicea*.

As might be expected, transitional types can be found among the substrates described above. Soils with varying textures and proportions of sand and silt differ in their water and salt regimes. Similarly, variation in the type and extent of rock outcrops and stones will lead to differences in soil characteristics.

SALT REGIME OF DESERT SOILS

The salt regime of the soil types described above is largely a function of the water regime (136). Salts carried from the ocean by rainwater are deposited and accumulate in fine-grained soils to the depth of the mean annual water penetration. Thus, soils which develop on loess, chalk, marl and clay are high in salt. The amount of accumulated salt depends on rock type and local climate. Some soils that develop on fissured rocks are salty, while others are not. Sands, gravels, and soil pockets of smooth-faced rock outcrops are effectively leached and thus not subject to salt accumulation.

SALT MARSHES

Salt marshes of varying size and salinity occur in coastal areas of Sinai, along the Dead Sea and in the 'Arava Valley. The marshes and zones of saline soil surrounding fresh water springs are formed by continuous wetting of the soil and subsequent evaporation. The shallow ground water is pulled upward by capillary action and substantial amounts of salt accumulate at the surface.

PHYSIOGRAPHIC INFLUENCE

Relief and geomorphology have a great influence on desert water regimes. Water infiltrates in varying amounts and depths depending on physiography so that hilltops, slopes and plains differ considerably in their water and salt regimes (123). The angle between the hill slope and the inclination of rock strata influences weathering, and thereby the hydrological characteristics of the slope. The bedded limestone in desert anticlines have few outcrops where the strata are nearly horizontal. Here precipitation is distributed evenly over the slope and no one site receives sufficient water to support more than a sparse vegetation. The same rock layer when inclined at a dip of 10° to 30° results in rock outcrops and a step-like contour when the hill slope is parallel to the rock layers. Runoff here is mainly diverted to the many fissures and these sites provide a good habitat for plants. At an inclination of 45° these strata result in little debris cover and the rock outcrops may be smoothened by the weathering activity of endolithic algae. Here, runoff flows into the few available fissures which comprise very favorable habitats for plants.

MAN'S INFLUENCE
HISTORY OF HUMAN ACTIVITY IN THE NEGEV AND SINAI

There is evidence today that human communities of hunters and gatherers lived in the Negev and Sinai from the Lower Paleolithic period, at least 300,000 years ago, up until the Neolithic period, about the 8th millenium B.C.E. The Paleolithic communities inhabited mainly the Negev Highlands, whereas the Neolithic people lived mainly in valleys and plains. Unlike today, we would surmise that these early inhabitants had only a negligible impact on the natural vegetation of the region.

The Epipalaeolithic and Neolithic periods (15th to 4th millennia B.C.E.) witnessed a major change in human culture and land use in the Middle East. Agriculture was developed in the region as evidenced by the remains of domesticated cereals and legumes found in archeological sites (141). At Jericho, in the Lower Jordan Valley, such crops were raised using spring water.

During the Chalcolithic period (4000 to 3100 B.C.E.), agriculture developed further with the establishment of large permanent farming communities. Many fruit trees were also domesticated at this time (142). Remains of domesticated plants (wheat, barley, pulses) as well as animals (sheep, cows and goats) at several archeological sites near Be'er Sheva indicate that the semidesert of the northern Negev supported an organized and established agriculture. Because the inhabitants lived in caves, they probably caused less destruction of trees than later populations that used wood for the construction of housing. However, copper mines at Timna and Punon depended on the use of natural woody vegetation from the 'Arava Valley as fuel for smelting operations.

Agriculture of the Early Bronze period (3100 to 2200 B.C.E.) was similar to that of the Chalcolithic. During this time, however, houses were built of stone and wood and incorporated the trunks of native trees in their construction. In addition, evidence of hundreds of small nomadic settlements dating from 2200 to 1950 B.C.E. has been found in the Negev. It is likely that these nomads exerted considerable pressure on the vegetation through grazing and by cutting woody plants for fuel. The copper mines of the 'Arava ceased operation during this period, thereby eliminating this use of trees. There is no evidence of substantial human activity in the region during the nine centuries starting at about 1950 B.C.E.

From the Iron Age and up to the end of the Byzantine periods (1200 B.C.E. to 640 C.E.) revitalization of human settlement in the Negev with the establishment of cities and farms took place (38). The Nabateans cultivated the valleys of the Central Negev Highlands using sophisticated techniques for the diversion and control of runoff water from winter rains. Grazing by nomads also continued in the region during this period. A steep decline in population and in agriculture took place in the Negev following the Arab conquest in 640 C.E. For more than 12 centuries, until about 1900, the Negev was dominated by a nomadic population whose main livelihood, the husbandry of goats, was extremely destructive to the natural vegetation. Beginning in 1900 the Ottoman government encouraged permanent agriculture among the Bedouin of the northern Negev and developed Be'er Sheva as a regional center. In Sinai, nomads have had an important influence on the vegetation up to the present day. Their destructive impact on vegetation is displayed in Figure 8.

With the establishment of the first Israeli agricultural settlements in the Negev in the 1940's, increasingly larger areas have been protected from overgrazing. Efforts have also begun to prevent soil erosion and to preserve valuable natural areas. However, the increase in population, the introduction of modern technology and the development of urban communities in the region especially since 1950 have resulted in substantial changes in the natural ecosystems. Wars and military training exercises have also had a destructive impact on the desert environment in large areas in the Negev and Sinai. Tanks and other armored vehicles, in particular, result in the destruction of plants and the disturbance of plant habitats. Ecologists and planners of the region face an important challenge in ensuring that the unique and productive value of the natural plant communities are preserved for future generations.

CUTTING AND GRAZING

The Bedouin burn lignified plants as heating and cooking fuel, a practice that often leads to the disappearance of woody plants in the vicinity of their encampments. The cumulative effect of Bedouin pressure can be seen in the sandy areas of northern Sinai (see discussion in District 8). When these nomads move to a new area, the pressure on the vegetation is released and the area is eventually repopulated with the original species.

Sustained overgrazing over the millennia may have resulted in permanent changes in the dominant vegetation. However, such changes are difficult to document owing to the absence of "control areas" not subject to grazing. We have no tools for analyzing the long-term influence of grazing and cutting since the Neolithic or Chalcolithic periods, but we have some short-term observations. Grazing leads to diminution in the size of edible plants and to temporary changes in their relative abundance, but not to the total extermination of species. During the few years that several Negev and Sinai areas were closed to Bedouin and their domestic animals, no substantial changes in the list of species and plant communities could be discerned. The prolonged cutting of lignified semishrubs in the hills near Be'er Sheva has led to dominance of *Asphodelus microcarpus,** which is neither lignified nor eaten by domestic animals (Fig. 6).

Fig. 6. The clumps of green leaves of *Asphodelus microcarpus* remain intact as the sheep graze around them in the hills near Be'er Sheva.

* For authors and common English names see index.

CULTIVATION AND FARMING

The Bedouin largely practice dry farming on loess soils. Shallow plowing disturbs only the surface crust, thereby increasing the roughness of the soil surface and reducing runoff. Such areas have a more favorable water regime and support a larger phytomass than areas not plowed. With the improved water supply weeds such as *Achillea santolina* and *Hyoscyamus reticulatus*, which are rare in other similar habitats, grow successfully.

In non-irrigated modern agriculture fields are plowed deeply and are supplied with manure and chemical nutrients. Because the ecological conditions here differ from those found in the Bedouin fields, the weeds are different as well.

Irrigated fields support weeds which are not found in other desert habitats. Included are *Conyza bonariensis*, *C. canadensis*, *Portulaca oleracea* and several species of *Amaranthus*. The actual flora of weeds in these fields is much influenced by the kind of herbicide treatment. The irrigated fields in the southern Sinai support a rich flora which was discussed by Hadidi et al., (1970).

NEW HABITATS AND INTRODUCED SPECIES

Human activity results in the creation of new habitats, such as roadsides and piles of construction debris. The natural vegetation is destroyed and those plants which succeed enjoy the absence of competition for water and nutrients. The removal of leached upper soil layers during road construction lead to the exposure of deeper saline soil. Such areas become populated with native salt-resistant species such as *Salsola inermis*, *S. volkensii*, *S. jordanicola*, *Aellenia hierochuntica* and *Atriplex leucoclada*. Such habitats may also support plants which arrive from distant regions which have no connection with the local flora. Such species are named Xénophytes (53) from the Greek xénos — stranger and phytos — plant, and are transported primarily by the actions of man.

Senniella spongiosa, introduced from Australia by the Institute for Arid Zone Research in Be'er Sheva as an experimental pasture plant, now grows along the roadsides near 'Arad, Dimona, the Dead Sea and Elat. *Kochia brevifolia*, introduced in the same way, is found along the roadsides in Be'er Sheva, 'Arad and Dimona. *Kochia indica* arrived in the Negev from India via Australia and then to Lower Egypt, where during the Second World War it was used as a pasture plant (94). The branched stems of *K. indica* were employed by the Egyptians during the 1948 war to pack their guns and ammunition. This species is common today throughout the Negev, and in waste grounds of El Qantara and along roadsides near the Suez Canal. *Aster subulatus*, introduced accidentally from North America in the 1960's, is prominent in irrigated fields of summer crops and along roadsides (24). Each plant produces thousands of achenes each with a pappus which allows them to be easily dispersed by wind.

Piles of household wastes and abandoned animal corrals result in areas rich in nitrogen which support specific flora for many years after the natural equilibrium is destroyed.

Mesembryanthemum nodiflorum and *Spergularia diandra*, which usually inhabit saline soils of the northern Dead Sea Valley, prosper as nitrophytes in these waste areas.

Although these new habitats are very conspicuous, the areas they cover are a relatively small part of the desert.

ANIMAL ACTIVITY

The impact of animals on plant life in deserts is complex and will only be discussed briefly.

Many desert plants are pollinated by insects (75). Some desert species are also pollinated by birds. The red-flowered mistletoe, *Loranthus acaciae* is visited by the Sunbird (*Cinnyris oseae*). It is suspected that other red-flowered species such as *Capparis decidua* are bird-pollinated as well.

The fruits of many species are adapted to seed dispersal by animals. The spiny fruits of *Emex spinosa*, *Onobrychis crista-galli*, *Tribulus terrestris* and *Neurada procumbens* become attached to large animals. The fruit of *Commicarpus* and of *Boerhavia* have mucilaginous glands capable of sticking to passing animals.

Harvester ants and species of the rodent *Acomys* collect many kinds of fruit, among them the hard-coated multiseeded fruits of *Medicago laciniata*, *Trigonella arabica*, *Limonium thouinii*, *Pteranthus dichotomus* and *Onobrychis crista-galli*. Many of the collected seeds are not eaten and are left to germinate in an arc or circle surrounding the nests. The multiple seed characteristic is apparently an adaptation to the unpredictable and unstable climate of the desert, and not to dispersal by ants and rodents (see p. 33). This has been shown in germination experiments in *Aegilops* (32) and *Pteranthus dichotomus* (38).

The evolution of the seed dispersal mechanism in *Acacia* is related to the life cycle both of bruchid seed beetles and of large herbivorous animals (83). In the 'Arava Valley and the Sarengeti Reserve of Tanzania, *Acacia* seeds are dispersed by large herbivores including gazelle, ibex, goat, rock-hyrax and elephant, which eat the fruit and excrete or spit the seeds. Intestinal passage increases the likelihood of germination of the hard-coated seeds (61). Herbivore digestion also kills the larvae and young of the bruchid seed beetles which develop in the seed and eventually destroy it. If the seeds are not entirely eaten by the beetles, they may still germinate.

Several plant species in the Negev and Sinai produce juicy fruits which are eaten by animals. These include *Capparis decidua*, *C. aegyptiaca* and *C. cartilaginea* in which the pungent seeds are surrounded by sweet pulp. The seeds and pulp are eaten by birds which transport the ingested seeds. The juicy fruits of *Lycium* spp., *Ochradenus baccatus*, *Ficus pseudosycomorus*, *Salvadora persica*, *Cordia sinensis*, *Ziziphus spina-christi*, *Balanites aegyptiaca* and *Phoenix dactylifera* (date palm) are transported by various animals.

The mucilaginous seeds of the parasite *Loranthus acaciae* are transported by birds. They remove the unwanted seeds from their beaks onto tree stems, where the seeds germinate.

Seed dispersal by birds has resulted in the recent introduction of several plant species at the Yeroham reservoir in the Northern Negev Highlands. *Potentilla supina* and *Cotula anthemoides*, both new introductions to Israel, and *Gnaphalium luteo-album,* a rare species of the Jordan Valley, now grow in the muddy soil along the reservoir in summer. The Yeroham reservoir is a migratory resting place for birds coming from Egypt where all three of these plant species thrive.

Animals affect desert plants in a number of other ways. Herbivorous animals such as gazelles, ibexes, camels, and goats, graze on desert plants. Larvae of various insects also feed on such plants. In some years, epidemic attacks of butterfly caterpillars prevent all annuals from setting seed. Leaves of *Atriplex halimus* shrubs and green stems of the semishrubs *Anabasis articulata* and *Anabasis syriaca* are the main food source of the sand-rat *(Psammomys obesus)*. Those shrubs under which they dig their holes are overgrazed most of the year and fail to set seed. The sand rat is highly susceptible to diabetes. Leaves of *A. halimus* were found to have a significant hypoglycaemic effect on these and other diabetic rats (2). Porcupines feed in summer on corms of *Colchicum tunicatum* and bulbs of *Tulipa amblyophylla* and *Scilla hanburyi*, decreasing the populations of these species each year. Soil erosion is increased at the sites where these geophytes are dug by porcupines (138), thus influencing other plants as well.

Many plants produce compounds which reduce or eliminate browsing by herbivorous animals. Such compounds, synthesized by many species at the onset of blooming and fruit setting, render the plant non-palatable. *Atriplex halimus* is one such species that is not equally flavorful throughout the year. *Stipa capensis* is grazed by goats only when the plants are young and again after seed dispersal. The accumulation of compounds which make the plant non-palatable during this critical stage of its life enables the species to complete its life cycle. The high proportion of odorous and poisonous desert flora may be an adaptation to predation by herbivorous animals. Species of the genus *Hyoscyamus*, including *H. boveanus* and *H. muticus,* contain high quantities of the alkaloids hyoscyamine, atropine, and hyoscine, and are frequently the only plants which remain in overgrazed areas.

Alkaloids produced by *Heliotropium rotundifolium, H. maris-mortui, Crotalaria aegyptiaca* and *Senecio desfontanei* (Schoental and Shani, unpublished) repel herbivorous animals. These compounds affect the liver of experimental animals and cause their death. A more detailed study was carried out on related African species which were found to contain hepatotoxic and hepatocarcinogenic pyrrolizidine alkaloids (112).

MUTUAL INFLUENCE OF PLANTS

Plants may influence each other in a variety of ways. The symbiotic association of fungi and seed plants (Spermatophyta) is known as mycorrhiza.

Seedlings of a European species of *Helianthemum* require a mycorrhizal fungus for the branching of roots and the emergence of the young stem. Without the proper fungus the seed will not develop. The fungus is only found with its symbiont host. The expensive truffle fungus has as its symbionts *Helianthemum sessiliflorum* and *H. ledifolium,* which occur in the Negev.

Tamarisk trees absorb deep saline water and excrete the excess salts through glands on their green stems and small leaves. The salt falling from the tree accumulates at the soil surface and creates a favorable environment for halophytes such as *Salsola inermis, Kochia indica,* and *Bassia muricata* (85). Other halophytes (e.g., *Reaumuria, Suaeda,* and *Atriplex*) shed their salt-excreting stems and leaves, creating saline patches in which only their own seedlings can grow (27).

Plants of *Stipagrostis scoparia* germinate and establish themselves in mobile sand dunes and thereby arrest dune movement. Thus is created an adequate habitat for the germination and establishment of *Artemisia monosperma*. This further promotes sand stability and prevents wing erosion of fine-grained silt. More water is now retained in the soil and many additional species take hold. *Stipagrostis*, which requires a continuous cover of new-blown sand, dies off after having prepared the way for the other species. This is a classic example of plant succession in which pioneer species create suitable conditions for other plants which replace them.

In sand deserts blue-green algae and fungi appear only when the soil crust contains at least 4-5 percent clay and silt. The algae and fungi increase soil stability by forming a surface crust. In time, this crust becomes thicker and harder to break. The crust creates new microhabitats and influences seed penetration (26). A similar biological crust on the soil of the regs decreases water penetration and contributes to runoff.

On north-facing slopes of hard limestone outcrops, crustose lichens smoothen the rock surfaces. This causes more water runoff to reach soil pockets which support rare plants in the desert.

RELATIVE IMPORTANCE OF ENVIRONMENTAL FACTORS

After having discussed the environmental factors which influence the distribution of plants in desert areas, we shall now compare their relative importance. This will help in choosing appropriate criteria for dividing the Negev and Sinai into subregions. Each subregion should be as homogeneous as possible and should differ substantially from the others.

According to climatic conditions the area under study can be divided as follows: 1) areas which support semishrubs on the entire slope; 2) areas which support semishrubs only in wadis (Fig. 14 and pp. 21-23). Each of these two climatic subdivisions may be divided by rock and soil type into many homogeneous units, each of which is associated with a plant association dominated by one or two species. The plant associations are therefore determined by the combination of climate and rock type.

The mutual influence of plants as a factor influencing the distribution of vegetation is negligible as compared with climate and soil type.

The impact of man and his domestic animals is generally local, but may be considerable in a few places. Such a site is the area around the boundary between the overgrazed sands of northern Sinai and the non-grazed sands of the Negev (Fig. 8). However, in general, the desert areas of Sinai and the Negev are less influenced than the more humid areas with denser human population. Therefore, we cannot divide the desert into subregions by using the criterion of man's influence.

Animals affect the vegetation over the entire area and have little influence on the distribution of plant communities.

To conclude, the two factors which clearly influence the distribution of plants and plant communities are soil and climate. We have, therefore, divided the desert into 19 areas which are here termed districts, based on geomorphological, climatic, and edaphic factors (Fig. 7).

DISTRICTS OF THE ISRAELI AND SINAI DESERTS; A GEOMORPHOLOGICAL DESCRIPTION

DISTRICT 1. JUDEAN DESERT

The Judean Desert descends from elevations of 600 to 700m. in the west to sea level in the vicinity of the fault escarpment bordering the Dead Sea. The descent extends along 16 to 20 km. The area consists of several small anticlines and synclines consisting of Senonian chalk and flint outcrops. Limestone and dolomite of the Cenoman-Turon era are exposed in the canyons and near the fault escarpment facing the Dead Sea.

DISTRICT 2. DEAD SEA VALLEY

This district is bounded on the west by the fault escarpment and to the south by the Lisan Marl cliff south of Sodom. The springs of the Sodom salt marshes originate at the foot of the Lisan cliff. The 'Arava Valley is to the south of these springs. The runoff water from much of the Galilee, Golan, Gilad, Edom, Moav, Judean Desert, Negev and eastern Sinai drains into the Dead Sea Valley (93). The valley is covered by young sedimentary rocks which were deposited in the "Lisan Lake". This lake preceded and was much larger than the Dead Sea (7). The "Lisan Marl" forms the peninsula shaped like a tongue (= Lisan in Arabic) of the Dead Sea. Pure marl was deposited in the center of the Lisan Lake while a mixture of conglomerate and marl was deposited along the shores. Springs are often found along fault lines and in alluvial fans of large wadis. Constant evaporation of the water supplied by these springs results in the accumulation of salts in the soil. Salt marshes have also developed in these areas.

DISTRICT 3. NORTHERN NEGEV HIGHLANDS

Asymmetric folds with steep southeast escarpments form the four anticlinal ridges of this district. The parallel axes of the ridges are oriented in a NE-SW direction. The SE escarpment slopes are steep with layers inclined to 90° or more while the

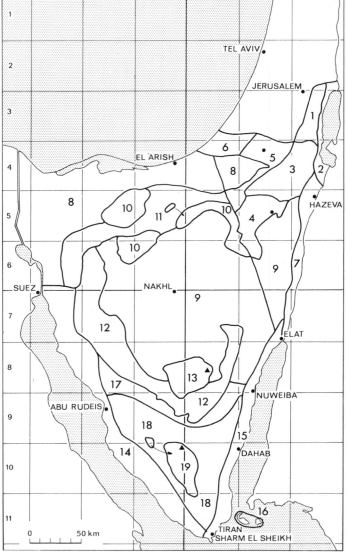

Fig. 7. Geomorphological districts of the deserts in Sinai and Israel.

northwestern slopes have a dip of only 5° to 10°. Upper Cretaceous limestones and dolomites are outcropped along the anticline crest and the escarpments. The weathering of this bedded and massive rock depends on the inclination of the strata (see p. 17). The youngest strata are exposed at the sides of the anticlines and are composed of Senonian flint and chalk. In the erosion craters of the Hatira and Hazera ridges, outcrops of Lower Cretaceous and Jurassic sandstones occur. The synclinal valleys consist of soft Senonian and Eocene chalks, marls and clays. The greater part of the area is covered by eolian and alluvial loess deposits derived from the adjacent ridges. Neogene sandstone and conglomerates make up the Yamin-Rotem and Yehoram-Dimona synclines. Weathered material derived from sandstone covers the consolidated rock. Picturesque badlands formed from weathered Tertiary marl and chalk occur in the Zin Valley forming the southern boundary of this district and the northern boundary of the Central Negev Highlands.

DISTRICT 4. CENTRAL NEGEV HIGHLANDS

This district is primarily an unfolded, horizontally bedded plateau composed of Middle Eocene and Turon inter-bedded limestone and chalk. Massive limestone strata form cliffs and smooth-faced outcrops near the erosion crater of Makhtesh Ramon and near Har Loz. Loess terraces in the valleys are famous for their support of ancient agriculture (37). The ancient inhabitants raised such Mediterranean crops as almonds, olives and grapes by diverting winter flood waters into terraced valleys of this otherwise dry soil.

The erosion crater of Makhtesh Ramon, located in the southeastern part of this district, contains rock outcrops not found elsewhere in this district. These rocks include Lower Cretaceous and Jurassic sandstone, gypsum, various clays, basalt and the plutonic crystalline rocks — bostonite and nordmarkite.

DISTRICT 5. NEGEV LOWLANDS

This district is made up of horizontally bedded Lower Eocene hard chalk in a landscape of mild hilly topography. Deep loess deposits of eolian origin have accumulated in the wide valleys of Be'er Sheva-'Arad, Nahal Sekher, and Shivta-Mashabbe Sade Valley. Mediterranean sands influence the soil texture. These blown sands are prominent near the belt of stony hills 13 km south of Be'er Sheva. Loess plains with occasional hills of Eocene rocks are typical of the Negev Coastal Plain and there is no precise boundary between the two districts.

DISTRICT 6. NEGEV COASTAL PLAIN

Cultivated loess plains are the most prominent feature of this district. Moving westward the texture of the loess gradually becomes coarse, ending with a belt of mobile dunes consisting of pure sand near the coast (Fig. 8). The channel of Nahal Besor, on its way to the sea from the Northern Negev, has cut through the loess of this area. It has formed large badlands which surround the water course. Clays, conglomerates, sandstones, and sands are exposed in the badlands, resulting in a unique mosaic of habitats.

Mobile sand dunes, which do not support agriculture and have sparse and overgrazed vegetation, appear white in aerial photographs when compared with the adjacent cultivated and vegetated sand fields (Fig. 8). Date palms, which use the fresh water from the high water table, create a dark belt between the sand dunes and the sea.

DISTRICT 7. 'ARAVA VALLEY

The 'Arava is a rift valley separated from neighboring districts by the prominent steep escarpments of the Edom Mountains and by the hills of the southern Negev. Most of the 'Arava Valley is composed of recent alluvial fans. Gravel plains are the dominant land form. Local water regimes are influenced by water table depth, soil properties, and wadi size. Tertiary sandstones, conglomerates, clay, and lacustrine

limestone are exposed between Hazeva and Nahal Paran. The high water table as well as winter floods contribute water and salts to the temporary salt marshes of Yotvata, 'Evrona and Elat.

DISTRICT 8. MEDITERRANEAN SANDS AND SALT MARSHES

Coarse-textured stable sands characterize the area in the west near the Suez Canal, whereas fine sand occurs in the Haluza sands to the east. In the entire area mobile dunes of varying density are scattered throughout the sandy undulating plain. The LANDSAT imagery (Fig. 8) shows the overgrazing of natural vegetation in northern Sinai west of the cease-fire line of June 1967. The contrast between the light-colored sands SE of the line and the dark-colored vegetated sands is a prime

Fig. 8. LANDSAT imagery of El 'Arish—Nizzana area as it apeared in 1973: 1. Sand, vegetation destroyed. 2. Sand, vegetation intact. 3. Date palm plantation on coastal sands. 4. Shifting Sands. 5. Sandy area covered by orchards of southern part of Gaza strip. 6. Fenced area of Sadot and Netiv haAsara where destruction has ceased. (Courtesy NASA).

example of human impact on vegetation (see also Otterman et al., 104). Date palms northeast of El 'Arish end abruptly to the east and give way to mobile sand dunes which, in turn, grade into vegetated loessial sands. Salt marshes abound in depressions and in coastal areas west of El 'Arish. The high water table brings salt from underground layers to the surface where it is concentrated by continuous evaporation. Bedouin collect sodium chloride from some of these salt marshes during summer.

DISTRICT 9. GRAVELLY PLAINS OF CENTRAL SINAI AND SOUTHERN NEGEV

Gravel plains have been formed over large segments of this district by the weathering of Senonian and Eocene chalk and chert and by the deposition of Quaternary alluvial material. The soils are of the reg type (15). Except in wadi channels where flood water leaches the salts, the ground under the gravel mantle is highly saline. As a result of the horizontal bedding and level topography there are relatively large areas which are edaphically homogeneous. Several small anticlinal ridges and limestone hills such as Gebel Kharim and Gebel Minshera are included here.

DISTRICT 10. TRANSITION BETWEEN DISTRICTS 8 AND 9

Shifting sands derived from district 8 cover these gravel plains which are at the northern range of district 9. The combination of sand (with its high permeability) covering fine-grained reg soil (with its high water storage capacity) produces habitats in which the water regime is more favorable than either sand or reg soil by itself.

DISTRICT 11. ANTICLINES OF NORTHERN SINAI

This district includes several folds of the Cenoman-Turon era, with limestone, chalk, dolomite, and marl outcrops. Extensive erosion in Gebel Maghara has exposed a sequence of 2000 m of Jurassic limestone, shales and sandstone. Large outcrops of smooth-faced limestone and dolomite are found at Gebel Halal, Gebel Maghara, Gebel Libni and Gebel Yiallaq. The syncline valleys are filled with sand-covered alluvium. The boundary between the southeastern sector of this district and the Central Negev Highlands is determined by elevation and precipitation.

DISTRICT 12. TABLE MOUNTAINS OF CENTRAL AND WESTERN SINAI

Most of this district consists of unfolded hills of limestone, dolomite, marl, and chalk formed in the Cenoman, Turon, and Eocene eras. Gebel et Tih in the central part of this mountainous belt has slightly inclined strata which weather almost like horizontal strata. Anticlinal structures and domes situated in proximity to faults with horizontal movements are common in the Mitla Pass, Gebel Somar, and Gebel Mineidra el Kebira (6). Small outcrops of smooth-faced limestone occur in the flanks of the folded structures and in the wadis at high elevations of Gebel et Tih. Many springs, including 'Ein Sudr, Moyet el Gulat, 'Ein Shallala, and 'Ein Abu Ntegina, occur in canyons draining the mountains.

DISTRICT 13. GEBEL EL 'IGMA

This district is a non-folded plateau of Eocene chalk and marl. It rises from an elevation of 500-600 m in the north, where it emerges from the gravel plains of District 9, to a peak elevation of 1600 m. Steep escarpments of marl, shales and chalk terminate the plateau on its southern, eastern, and western sides (Fig. 51). The area above 1200 m has smoother topography than the lower areas as a result of deeper soil accumulation.

DISTRICT 14. COASTAL PLAIN OF THE GULF OF SUEZ

This district is characterized by young Tertiary and Quaternary sediments of alluvium, sandstone, gypsum, and limestone. Much of the alluvium comes from the hills to the east. Alluvial fans derived from magmatic and metamorphic parent rocks are common in the southern part of the district which drains the southern Sinai Massif. Limestone dominates along the border with District 12. Throughout District 14 sand, derived from local weathering and transported by wind, mixes with the alluvium or forms individual dunes. The abundance of dunes increases northwards towards the sand fields of District 8.

DISTRICT 15. COASTAL PLAIN AND FOOTHILLS OF THE GULF OF ELAT

The eastern foothills of the Sinai, descending sharply towards the coast, are composed of metamorphic and magmatic rock. The southern third of the coastal plain is broad, whereas the northern part is narrow. Alluvial fans derived from magmatic and metamorphic rock cover most of the coastal plain. From south of Ras Nasrani to Ras Muhammad, a large area near the beach is built up of fossil coral reefs. Wind-blown sand accumulates in wadis dissecting the coral reefs.

DISTRICT 16. TIRAN AND SINAFIR ISLANDS

These two islands are composed of Tertiary and Quaternary gypsum, sandstone, fossil coral reefs, and alluvium. Tiran Island is more diverse topographically and edaphically than Sinafir Island. Wind-blown sand accumulates in depressions and wadis, and also mixes with the alluvial material near the shores.

DISTRICT 17. SANDSTONE BELT

Nubian sandstone, sand derived from this sandstone and some basalt outcrops, are the main edaphic elements of this district. Massive sandstone forms smooth-faced outcrops mainly near the highest part of the district at the foot of Gebel Dalal. The most common rock formation consists of layers of

Plate 2.   Smooth-faced outcrops of red granite supporting *Pistacia khinjuk* and many other rare plants at the top of Gebel Serbal.

hard sandstone interbedded with softer layers containing silt and clay. Several hilltops, such as Gebel Sarabit el Khadem, are made up of basalt rocks interbedded in the above-mentioned sandstone formation. Sand derived from weathering of the sandstone accumulates in valleys and forms sand fields in much of the district.

DISTRICT 18. LOWER SINAI MASSIF

The elevation of this district is between 500 m and 1500 m. Several types of granite, other igneous, volcanic and metamorphic rocks are the main rock types of this district (40). Alluvium derived from them accumulates in the wide valleys. Small areas of smooth-faced granite occur, and coarse talus covers the slopes. Fine-grained soils are rare in this district.

DISTRICT 19. UPPER SINAI MASSIF

Smooth-faced granite outcrops are prominent in this district, forming mountains such as Gebel Serbal, Gebel Umm Shaumar, Ras Safsafa and Gebel Bab. Black mountains consisting of old volcanic rocks are rather common, the most prominent of which are Gebel Katherina and Gebel Abbas Basha. The soil accumulated on the slopes and the talus contains a higher percentage of silt and clay than in District 18. Fine-grained soils occur in old wadi beds. Hundreds of springs and wells support a unique agriculture consisting of many small gardens throughout the district.

CHAPTER 2. DESERT ADAPTATIONS OF PLANTS

LIFE FORMS AND THEIR SIGNIFICANCE

The extent of vegetative cover in the desert varies seasonally and annually according to the supply of available water. Extensive carpets of colorful annuals may cover an area during the spring of a rainy year. The same area will be devoid of annuals in a dry year. In order to analyze and describe these changes in vegetation, five principal life forms of seed plants are recognized. These life forms are based on the location of renewal buds and on which plant parts are shed during the season unfavorable for growth (109). The five life forms considered here are:

1. Therophytes — annual plants. These have the renewal bud in their seed. During the dry season all the other parts of the plant are dead (Fig. 9).

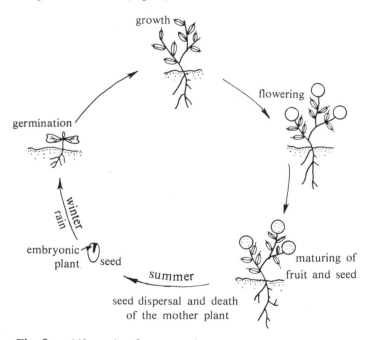

Fig. 9. Life cycle of an annual.

2. Geophytes — perennial plants having the renewal bud under the soil. The buds are protected by dry leaves and sustained by food reserves in the storage organ. During summer, above-ground parts of geophytes desiccate and are shed (Fig. 10).

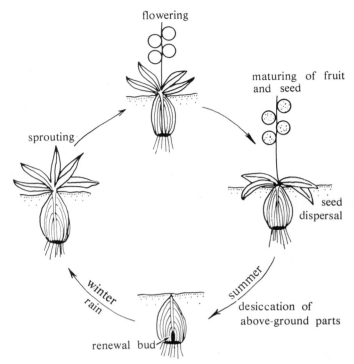

Fig. 10. Life cycle of a geophyte.

3. Hemicryptophytes — perennial plants in which the renewal bud is located at the soil surface at the top of a storage root or system of thick roots. The above ground parts desiccate in summer and the dry leaves protect the bud (Fig. 11).

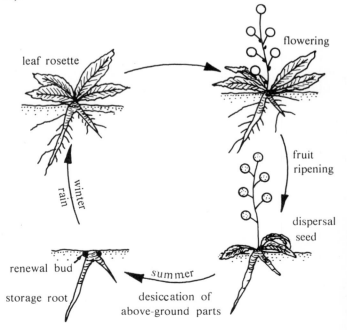

Fig. 11. Life cycle of a hemicryptophyte.

24

4. Chamaephytes — semishrubs. These have renewal buds above the ground. Branches become dry and are shed according to a predictable sequence. The renewal buds are usually located at the base of flower-bearing branches which become dry after seed dispersal. Hence, the dwarf size of the semishrubs is retained (Fig. 12).

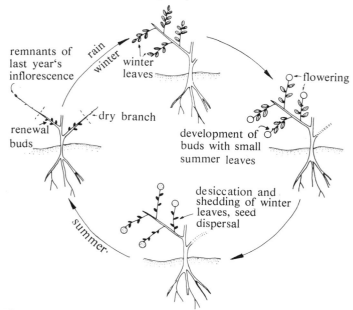

Fig. 12. Life cycle of a chamaephyte (semishrub).

5. Phanerophytes — shrubs and trees. Here the renewal buds are located at a considerable height above the ground. Branches are not shed. The leaves are replaced seasonally and there is no evident order in the death of branches. The extent of branch loss depends on the water regime (Fig. 13).

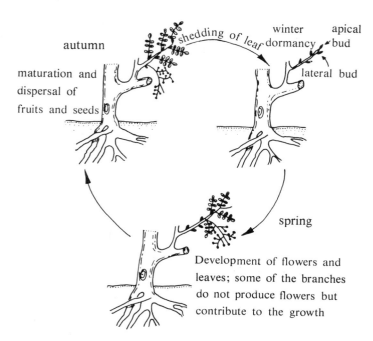

Fig. 13. Life cycle of a tree or shrub (phanerophyte).

In the first three life forms the plant is not visible in the unfavorable season because the above ground parts are shed. They were named "drought evaders" (95) or "arido-passive plants" (38).

Life forms 4 and 5 are prominent during the whole year and can be seen even in dry years; they were named "drought persistents" (95) or "arido-active plants" (38).

Ferns, mosses, liverworts, lichens, and algae, which are not seed plants, reproduce by spores. Under desert conditions the entire plant dehydrates during the dry season but does not die. It remains in an anabiotic state of desiccation and resumes activity following wetting. Such plants are also known as "poikilohydrics" (129). The ferns *Ceterach officinarum, Cheilanthes fragrans,* and *Cheilanthes catanensis* may also be included among these "resurrection plants".

Three major types of areas can be distinguished based on typical fluctuations in the populations of each of the five life forms of seed plants (Fig. 14):

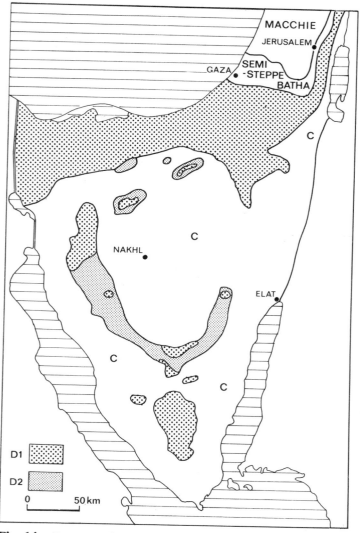

Fig. 14. Patterns of semishrub distribution in the deserts of Israel and Sinai: Dl—vegetation in a diffused pattern determined by climate; D2 — vegetation in a diffused pattern determined by soil and rock type; C — vegetation in a contracted pattern.

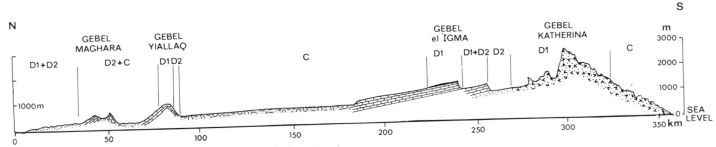

Fig. 15. A schematic north-south cross-section of Sinai showing semishrubs distribution (symbols as in Fig. 14).

1. Areas having an annual rainfall above 300 mm. All the life forms are present every year, but the occurrence of annual species varies annually according to rainfall. Trees are rare in the areas having less than 400 mm rainfall, and are restricted to wadis and to soil pockets in outcrops of smooth-faced rock. Populations of semishrubs are constant and provide relatively dense cover on stony slopes. Deep loess-like soils, having a poorer water regime than stony slopes, support relatively few semishrubs.

2. Areas having 80 to 250 mm annual rainfall. In these areas the number and type of annual plants vary considerably from year to year. In crevices and soil pockets of smooth-faced hard limestone and dolomites, annuals are present nearly every year. Smooth-faced granite outcrops in Sinai usually do not contain enough soil in their crevices to support annuals. The number of years in which annuals appear is greatest in stony and sandy soils, lower in the loessial soils, and still lower in the chalk and marl outcrops. Hemicryptophytes and geophytes may produce leaves during years in which annuals do not appear. At this range of precipitation, semishrubs occur in all stony soils and even on fine-grained soils in more humid areas within this range. In drier areas semishrubs occur on fine-grained soils only in wadis having sufficient moisture for their development. On some soils droughts may lead to the death of semishrubs. Shrubs and trees occur in large wadis that retain sufficient ground water to support phanerophytes in drought years. The phanerophytes typically occur in association with the smooth-faced rock outcrops.

3. Areas of less than 50 mm annual rainfall. A given site may be devoid of annuals for many years. Even in rainy years the annuals are restricted to the wadis except in areas of sandy and rocky soil. Geophytes and hemicryptophytes are generally sparse. In most soils, semishrubs are restricted to wadis. Semishrubs may die off in many sites during drought years. Sandy and rocky slopes support semishrubs both in and out of wadis. If there are several consecutive dry years even large trees found in wadis may die of desiccation. Such deaths are rare, and appear to result from a combination of drought years and erosional changes in the course of the wadi beds.

Climatic fluctuations from year to year cause drastic changes in the population of arido-passive plants (which shed their above-ground parts). Therefore, these plants are not useful in describing the basic structure of desert vegetation.

The arido-active plants constitute permanent elements of the vegetation and are more useful for describing plant cover in deserts.

Pattern of vegetation which is restricted to wadis was named by Monod (92) "mode contracté", whereas that which is found all over the area was called "mode diffus". We shall refer to these as contracted vegetation and diffused vegetation respectively. These patterns are influenced by water regime and therefore may be correlated with climatic and edaphic conditions. In areas with less than 50 mm annual rainfall, most of the vegetation is contracted. In areas with more than 80 mm of rainfall, most of the soils support diffused vegetation. In each of these areas there are exceptions caused by soil conditions. Fine-grained soils, such as those derived from clay, marl and loess, support contracted vegetation even in a climate which favors diffused vegetation. Diffused vegetation may be found in areas of 30-50 mm of rainfall in sandy and rocky soils because of their good water regime. We designate an area with diffuse semishrub vegetation which is climatically controlled as D_1 (Figs. 14 & 15) and that of diffused vegetation as a result of edaphic influence with D_2. The contracted vegetation in the area shown in Figures 14 & 15 is mainly climatically controlled and represented as C. The boundary between the climatically controlled types, D_1 and C, follow the isohyet of 70 mm.

Throughout the desert, springs and salt marshes, having a constant water supply, do not exhibit large fluctuations in plant cover. Because of the high soil salinity and poor aeration, these wet habitats principally support semishrubs, shrubs, and trees. In dry regions the vegetation pattern of springs and salt marshes should be considered as D_2.

ADAPTATIONS OF PLANTS

Environmental adaptations can be studied at several levels. These levels include: the population level; the entire plant; particular organs, such as stems, leaves, roots, or seeds; particular plant tissues; cells and their constituent organelles; and finally the biochemical processes which take place in the organelles. We shall restrict our discussion here mostly to the level of plant organs because adaptations of stems, leaves, and seeds can be easily observed in the field.

Xerophytes are plants generally adapted to life in areas where the water supply is limited. Xerophytes, in contrast with hydrophytes, exhibit only a few adaptive structural modifications. All plants in non-aquatic habitats must meet the challenge of drought, and the transition between mesophytes and xerophytes is not sharp. Maximov (89) points out that xerophytes exhibit so many mechanisms for resisting drought that it is nearly impossible to identify common features. In discussing the mechanisms regulating germination, Koller (78) notes that there is no mechanism that is peculiar to specific habitats, and there is no specific physiologic, anatomic, or morphologic property that occurs only in desert plants. The differences between mesophytes and xerophytes are quantitative rather than qualitative. Plants having adaptations to dry conditions can be found in both xeric and mesic environments. However, the proportion of plants showing such adaptations is much higher in deserts.

Xeromorphs are plants with specific structures adapted to retard transpiration. In areas with a typical Mediterranean climate, where summer rains are rare, xeromorphic plants are found even in relatively wet habitats. For example, *Nerium oleander,* which grows in Israel on river banks, has typically xeromorphic leaves.

ADAPTATIONS OF ABOVE-GROUND PARTS

ENVIRONMENTAL DAMAGE TO PLANT PARTS

Before examining xerophytic adaptations of stems and leaves, we must consider the kinds of environmental damage that plants must withstand. Desert plants must withstand scarcity of water, high quantities of salts, high temperatures, and high levels of radiation.

Most physiological processes depend on water. Water is the most important component of plant protoplasm and most tissues fail to function when dehydrated by 10 percent (13). Nutrients are dissolved in water and are absorbed by roots and transferred to plant tissues in aqueous solutions. Water is required for photosynthesis and for maintaining the structural firmness of the young plant parts. The high specific heat of water reduces the likelihood that plant tissues will become overheated. The loss of water through transpiration also helps to cool the plant. Reducing the water content below a certain threshold (which differs from plant to plant) will result in damage to the protoplasm and the cessation of essential life processes.

Temperature exerts an important control on physiological activity. Many processes will stop when the temperature falls below a certain threshold but will resume at its former rate when the temperature rises again. Physiological processes are accelerated with rising temperature, leading to the accumulation of waste material that can cause poisoning. At high temperatures many proteins disintegrate, including essential enzymes. Succulent plants can withstand very high temperatures, i.e., 58° − 65° (13), owing to the presence of organic acids in their cells. These acids neutralize poisonous ammonia produced by the disintegration of proteins by heat. The high water content of succulents may serve to dilute these acids and prevent them from damaging the plant.

Salt accumulates in many desert soils as a result of the continuous evaporation of water, either from rain or from springs. Some plant species can selectively absorb water and thereby prevent the entry of salts into their tissues. Other species utilize special glands to excrete excess salts (126). Still others, the succulents, dilute salts in their tissues with large quantities of water.

It is apparent that desert plants must be able to prevent excessive water loss and overheating. Water escapes from plants through the stomata, a process which also cools the leaves. However, the plant requires open stomata in order to absorb atmospheric carbon dioxide (CO_2) which is essential for food production. If the stomata open for CO_2 absorption, excessive water loss may occur. If, conversely, the stomata are closed to prevent excessive water loss, the plant may lack sufficient CO_2. Plants have evolved many solutions to this problem (89).

ADAPTATIONS OF LEAVES

In most seed plants leaves are the principal organs for photosynthesis and transpiration. According to Fahn (42), desert plants in general have smaller leaves, and leaf functions are often assumed by the stems. Many semishrubs and shrubs develop large leaves in winter, promoting efficient photosynthesis. In the dry season the large winter leaves are either shed or replaced by small summer leaves, reducing the transpiring area (101). Among the species which remain leafless in summer are: *Moricandia nitens, Asteriscus graveolens, Lycium* spp., *Rhus tripartita, Suaeda asphaltica, Chenolea arabica,* and *Euphorbia hierosolymitana.* Plants which develop small summer leaves include (see 102): *Artemisia herba-alba, Sarcopoterium spinosum, Reaumuria hirtella, Salvia dominica, Phlomis brachyodon* (Fig. 16), *Ph. platystegia, Ph. aurea,*

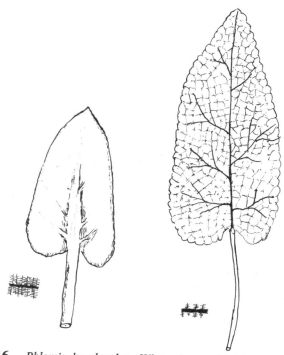

Fig. 16. *Phlomis brachyodon.* Winter leaves (right) showing an enlarged cross-section of the lamina below. Summer leaf (left).

Origanum dayi, O. isthmicum, O. ramonense (Fig. 100), *Anvillea garcinii* and many others. This adaptation is very common among Mediterranean semishrubs as well.

A reduction in the transpiring area may also be achieved by the folding of leaves so that the stomata face inwards. This mechanism is found in: *Helianthemum kahiricum, H. ventosum, H. sancti-antonii, H. sessiliflorum,* and *Fumana thymifolia* of the Cistaceae, as well as in many perennial grasses. *Halianthemum* and *Fumana* have small summer and large winter leaves as well. *Zygophyllum dumosum* (Fig. 66) sheds its leaflets in summer and retains only the petioles. This decreases the transpiring surface by 90 percent (99, 102).

Many desert species have smaller leaves than their relatives in more humid areas. However, a plant with many small leaves may have the same leaf surface area as a plant with a few large leaves. Measurements reveal that the leaf surface area of pine trees is similar to that of broad-leaved trees of the same size. The advantage of small leaves is the proximity of the photosynthesizing tissues to the vascular tissues.

An important property of desert plants is a well-developed vascular system. The density of veins is higher in leaves of desert plants than in mesophytic races of the same species (125). This adaptation enables plants to quickly counteract water losses from the leaf tissues. Extensive vascularization is exhibited in *Lactuca serriola*, a Mediterranean species that grows during the dry summer (Fig. 17). Squares were taken from *L. serriola* leaf blades and the total vein length per square was measured. It was found that near the leaf center the vein length was 1.6 mm per square millimeter of leaf surface, whereas near the margins it was 2.4 to 2.8 mm. Even though *L. serriola* has an efficient water transfer mechanism, many specimens have dry leaf margins. In *L. serriola,* upper leaves, i.e., those furthest from the roots, are the smallest and appear to be more xeromorphic than the lower leaves. The latter begin to develop during the humid season.

In the course of evolution, plant leaves have developed certain anatomical features which serve to reduce water loss. These include a multicellular epidermis, sunken stomata, a thick cuticle, and wax layers. In those mesophytes, in which the cuticle is thin, cuticular transpiration continues even after the stomata are closed. In xerophytes, having a thick cuticle or other leaf coating, water loss is negligible if the stomata are closed. Xerophytes may also open their stomata during hours of low water loss, photosynthesize efficiently, and then close the stomata under increasing water deficit. The transpiration rate in xerophytes not under water stress may exceed that of mesophytes.

The summer leaves of *Capparis aegyptiaca* are coated with wax giving them a lighter color than the winter leaves. The lighter color reflects solar radiation effectively so that the leaf is less likely to overheat. The hairs present on the leaves of many desert plants also increase reflectance and decrease cuticular transpiration. The effectiveness of white hair cover in reflecting sun radiation can be demonstrated in *Phagnalon*. If you use a magnifying glass to focus the sun's rays on dry white "wool"

Fig. 17. *Lactuca serriola* — vein system of leaves as seen in the morning light.

derived from *Phagnalon* stems (see p. 127), the radiation is reflected. The point of ray concentration looks bright and it takes a long time to burn. However, if the wool is darkened with a pen or pencil it will ignite almost immediately at the marked spot. Desert forms of European lichens have a much lighter color in the desert (47). In hirsute species that undergo leaf replacement, the leaves are much hairier in summer. Such hairy species are: *Phlomis* spp. (Fig.16), *Artemisia herba-alba, Anvillea garcinii, Helianthemum* spp., *Fumana* spp., *Majorana syriaca, Marrubium alysson,* and many others. *Paronychia argentea* and *P. sinaica* have translucent shiny white stipules which are longer than the summer leaves and hence protect them from excessive radiation.

Leaf and stem hairs also reduce water loss by trapping water vapor near the plant surface. The hairs reduce transpiration by making the leaf-to-air moisture gradient less steep.

Leaves of *Atriplex* have a different color in the wet and dry seasons. The leaf surface is covered with vesicular hairs containing a salty solution. In winter the vesicles are full and transparent. In summer when the water in the vesicle evaporates, they become dry, giving the leaf a light color which reflects the sun's rays. These vesicles, known in many species of *Atriplex,* are especially prominent in *A. leucoclada* and *A. halimus.*

The most obvious function of leaves is not to prevent loss of water but to supply the plant with the products of photosynthesis. Three pathways of photosynthesis are known today. (For further information see 9, 82 and 134). The most common is the C_3 pathway in which the first product of CO_2 fixation is a molecule with three carbon atoms. The C_4 pathway results in the formation of an organic acid with four carbon atoms. Many authors believe that CO_2 fixation is more efficient in C_4 plants than in C_3 plants, so that C_4 plants may photosynthesize efficiently during short periods of stomatal opening. C_4 plants are also better adapted to hot desert environments because they require high temperatures for optimal photosynthesis (30°–47°C versus 15°–25°C in C_3 plants). Of the desert plants from Israel and Sinai listed by Winter and Troughton (134) as C_4 plants, most are succulent xerohalophytic semishrubs of the Chenopodiaceae. These are: *Salsola inermis, S. tetrandra, Anabasis articulata, A. setifera, Bassia muricata, Aellenia lancifolia, Atriplex halimus, Chenolea arabica, Seidlitzia rosmarinus, Halogeton alopecuroides, Hammada scoparia, H. salicornica, Suaeda fruticosa, S. monica, Noaea mucronata, Haloxylon persicum, Calligonum comosum.* The last three are not confined to saline soils.

It appears that C_4 plants are especially well adapted to salty dry soils where they dominate the arido-active vegetation. This agrees with Leatsch's (82) hypothesis that some C_4 plants have evolved in hot regions with soils rich in cations.

The Crassulacean Acid Metabolism (CAM) pathway of photosynthesis was first described in the Crassulaceae, a family that includes many succulents. In CAM, CO_2 fixation takes place at night so that the stomata need not open during the day to take up CO_2. Winter (133) and Winter et al. (135) found that *Mesembryanthemum crystallinum, M. nodiflorum,* and *M. forsskalii* use CAM when the plants are under drought or salinity stress, but convert to the C_3 pathway when the stress is removed.

Leaves of many desert plants contain water storing tissue. In a succulent leaf of *Anabasis setifera,* for example, a thin layer of chlorenchyma (photosynthesizing tissue) is under the epidermis. Near the small central cylinder of the vascular system are large water storage cells. These cells can instantly replenish the water lost by the chlorenchyma cells. When losing water, the cell walls of the water storage tissue fold like an accordion and the cells shrivel up. When supplied with water they return to normal size. The petioles of *Zygophyllum dumosum* shrink and become wrinkled when dehydrated, but swell to two to three times their size two days after the first effective winter rain. The importance of this "reservoir", which can be depleted and refilled without damage, was apparent at the end of a dry growing season in the northern Negev. During a severe late spring drought, many plants died and all leaves of *Z. dumosum* were small and wrinkled. However, after a rain the leaves swelled and the plants completed their life cycle and set seeds. Hence, the water tissue enabled the plants to survive a period of severe drought.

ADAPTATIONS OF THE STEM

Stems contain some of the most important adaptations of plants to desert life (42). "Stem assimilants" are species which have green stems that carry the main burden of photosynthesizing throughout the year. Some of them have leaves for only a short period. This adaptation also occurs in some plants in Mediterranean habitats that are subject to hot, dry summers. *Retama raetam* is the most common stem assimilant in our desert. This plant develops new soft leafy stems during the rainy season. The small leaves are shed in spring, while the soft stems develop a thick epidermis and become rigid. However, the stems remain green and photosynthesis is carried out by the chlorenchyma situated along the longitudinal furrows (42). The stomata are located in these furrows below the stem surface and are also protected by hairs. Transpiration in *R. raetam* may be reduced in summer to 3 percent of winter levels (37). The thickly cutinized epidermal cell walls (87) virtually eliminate cuticular transpiration.

In other stem assimilants the outer epidermal walls rapidly swell when they absorb water, but release the water very slowly (87). These walls also reduce cuticular transpiration. This mechanism is characteristic of various desert stem assimilants including: *Anabasis articulata, A. setifera, Hammada salicornica, H. scoparia, Crotalaria aegyptiaca, Calligonum comosum,* and *Ochradenus baccatus.* Other stem assimilants whose epidermal structure has yet to be studied are: *Ephedra spp., Farsetia aegyptiaca, Zilla spinosa, Pituranthos tortuosus, P. triradiatus, Polygonum equisetiforme, Scrophularia xanthoglossa, Polygala sinaica,* and *Astragalus camelorum.*

Some perennial grasses behave as stem assimilants in summer when their leaves and upper stem internodes desiccate. Such grasses include: *Piptatherum miliaceaum, Panicum turgidum, Tricholaena teneriffae, Pennisetum divisum, Stipagrostis scoparia, S. lanata, S. raddiana, Hyparrhenia hirta,* and others.

Many people associate plant adaptation to deserts with the succulent cacti. The Cactaceae, which is essentially a family of the New World, have rudimentary leaves and thick green stems made up principally of water-storing tissue. Succulents resembling cacti occur in other plant families — Euphorbiaceae, Compositae, Crassulaceae and Asclepiadaceae. Many of these succulents use the CAM photosynthetic pathway. Under drought conditions they seal their stomata very efficiently and use mostly CO_2 produced through respiration for photosynthesis. *Caralluma* spp. of Israel and Sinai can withstand prolonged drought. Three branches of *C. negevensis* were collected in the northern Negev, 20 km east of Dimona. Their mean daily loss of water was 0.0013 to 0.0054 percent of their weight. Two branches died after eight months, but the third lived for 12 months (Fig. 18). The dead branches weighed 9.3 percent and 13.5 percent of their original weight. At the end of 12 months the surviving branch weighed 36.2 percent of its original weight. Some roots remained on the surviving stem and its wounds were probably smaller than those of the other two branches. In contrast to *Caralluma,* other desert plants have been found to lose water equivalent to their entire weight within an hour.

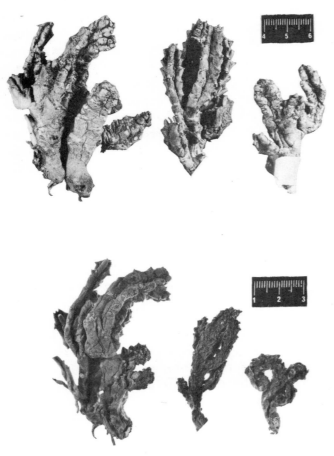

Fig. 18. Stems of *Caralluma negevensis* that were kept in a room, upper —July 17, 1974. Lower— July 17, 1975.

Another adaptation to drought is shown by *Phagnalon rupestre*. This plant sheds most of its leaves in summer and its stems and branches are covered with a white woolly bark. The bark protects the chlorenchyma from the dry air and from the overheating effects of the intense sunlight. *Phagnalon barbeyanum* and *Ph. nitidum* exhibit the same adaptation. *Moricandia nitens, Lactuca orientalis* (Fig. 19), *Euphorbia*

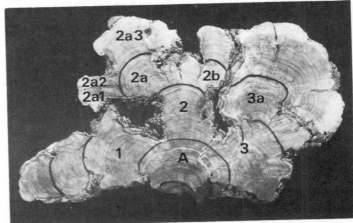

Fig. 19. *Lactuca orientalis* in summer when it has no leaves but produces flowers and fruits.

hierosolymitana, Gymnocarpos decander, and species of *Lycium* all shed their leaves in summer and have a shiny, light-colored bark. *Calligonum comosum* is a stem assimilant which has green stems in winter and lignified, light-colored, shiny stems in summer. It possesses "skeleton stems" which sprout in winter and lignify by summer. The ephemeral soft stems which carry the flowers and the fruits sprout during the rainy season and fall off the plant in summer. The date of stem shedding depends on the water regime in the particular habitat where each plant grows. Many semishrubs and trees possess living fibers that store water and food materials that may be used during dry periods (42). In this respect, these fibers function like the storage tissues of succulents, or the bulbs and corms of geophytes.

An interesting adaptation is exhibited by plants in which the trunks split into many units, each with its own root system (49). The trunks of *Artemisia herba-alba* contain rays of suberized cells that divide each plant into many independent units (Fig. 63). In a dry year, those units whose roots did not supply sufficient water will perish. Other units whose roots penetrated rock fissures and crevices where water was available will survive. The same adaptation is found in *Peganum harmala* and *Achillea fragrantissima.*

In *Zygophyllum dumosum* each main branch has its own root. Cambial activity is restricted in drought years to those branches whose roots have penetrated a moist microhabitat. The trunk of the plant is rarely circular in cross section and the annual growth resembles arches rather than rings. The trunk of a 100-year-old *Z. dumosum* from a wadi in the 'Arava Valley is shown in Fig. 20.

Fig. 20. A cross section in a trunk of *Zygophyllum dumosum* showing growth increments in different years: A=1867—1890; 1, 2, 3 = 1891—1915; 2a, 2b, 3a = 1916—1941; 2a1, 2a2, 2a3= 1942—1967.

This section started to grow in 1867. The shrub had one main trunk and root (A) which split into stems (and roots) 1, 2, and 3 during three consecutive dry years about 1890. Stems 1 and 3 expanded until 1915 when a two to three year drought resulted in partial cessation of cambial activity. The arches in the trunk became shorter, suggesting that the active crown was smaller. In 1915, stem 2 split into 2a and 2b. In 1941, following several dry years, 2a further divided into 2a1, 2a2, and 2a3. Discontinuity of growth arches, such as between 1 and 2, or 2a2 and 2a3, represent the cessation of cambial activity. Stem 3 did not divide in 1941, suggesting that the wadi changed course in such a way that root 3a still received sufficient water but root 2a did not. The narrow growth increments in 1890, 1915, and 1941 provide evidence that these were dry years. In each case, splitting occurred soon afterwards. The growth pattern we described in trunks and roots of *Zygophyllum dumosum* is also found in *Juniperus phoenicea.* Individuals of both species may survive hundreds of years.

ADAPTATION OF ROOTS

Root systems of many desert plants, especially of arido-active species, are more extensive than their above ground parts. In wadis and sandy soils, where water penetrates deeply, roots are particularly long. In sandy habitats, the roots of *Retama raetam, Artemisia monosperma,* or *Calligonum comosum* may be 10 to 15 m in length. The development of root systems in the desert often reflects the water regime of particular sites. Near Sede Boqer (annual rainfall 100 mm), water penetrates the loess soil down to 40-60 cm; the roots of the *Zygophyllum dumosum* growing here only reach this depth. Near Masada (annual rainfall 50 mm) the roots of the same species penetrate 1-3 meters into the fissured dolomite rocks reflecting the deep water penetration here.

The root system of *Z. dumosum* consists of perennial lignified roots with a thick bark. Scattered throughout the surface of these roots, there are nodes under the bark. Shortly after an effective rain, thin ephemeral roots develop at the nodes. These ephemeral roots rapidly absorb water (50) which is then transferred to the succulent leaf petioles where it is stored. The root systems of *Anabasis articulata, Noaea mucronata,* and other species function in a similar way. The development of ephemeral roots has also been observed during late summer and fall in ten arido-active species in the Negev and Sinai. This may be related to increasing amounts of dew during this season.

Roots of many species contain water and food reserves. The swollen roots of *Erodium hirtum, E. glaucophyllum, E. arborescens,* and *Asparagus stipularis* are clearly storage roots. Of the 19 species of *Erodium* growing in Israel and Sinai, only these three desert species are perennials.

Root systems of plants growing in sand subject to continuous wind erosion must be adapted to withstand desiccation. In sandy habitats of northern Sinai, the dominant plants, *Convolvulus lanatus* and *Artemisia monosperma,* have taproots protected by a thick bark.

ADAPTATIONS OF THE DISPERSAL UNIT

Dispersal unit is defined as the seed together with any additional part of the plant which functions in the dispersal of the seed. This additional part may be the whole fruit, sepals, and even the whole plant.

A seed is a dormant young plant covered by a seed coat which protects it from external hazards. This dormancy enables many individual plants to be protected from harsh environmental conditions, such as drought, extreme cold, and predation, and ensures the survival of the population through periods which are not suitable for growth. The dormant embryo protected by its seed coat demands very little from its environment. Its resistance to extreme drought and high temperature is greater than at any other stage in the life of the plant.

The mechanisms controlling where and when the seeds will germinate are the main survival strategies of many species in the desert. The time and place of germination influence the conditions of soil, water, nutrition, and microclimate that the adult plant will face in its lifetime. Seedlings of a species which germinates easily after the first rain may face a situation where after a short while many of the young plants will die from desiccation. The danger of a long dry season after a light or medium shower is much more probable in the desert than in moister regions. Many desert species have adapted to an unpredictable water supply by developing ways of gauging the amount of rainfall and thus preventing germination before sufficient water is assured. Three ways in which plants gauge water in order to ensure germination and seedling establishment are: 1. delay of seed dispersal until the rainy season; 2. delay of germination by chemical inhibitors; 3. delay of germination by low permeability of the seed coat to water.

Certain desert plants produce a seed population which is heterogeneous with regard to requirements for germination. Some of the seeds germinate in the first season while others remain dormant in the soil. In case of a natural catastrophe, such as drought or grasshopper infestation, the dormant seeds are available to germinate in future seasons.

SEED DISPERSAL MECHANISMS AS WATER GAUGES

Plants which disperse their seeds by raindrops are known as hygrochastic plants. The seeds of *Anastatica hierochuntica* and *Asteriscus pygmaeus* are enclosed for a long time within the mother plant where they are protected from predator animals. After they ripen, the seeds (achenes) of the composite *Asteriscus pygmaeus* are enclosed in heads covered by stiff involucral bracts. Five minutes after wetting, the bracts open (Fig. 126) and the achenes are exposed. Additional heavy rain may release some seeds from the heads. When the head becomes dry the bracts close and cover the remaining achenes. Heavier rain is required for releasing achenes than for merely opening the involucral bracts; both processes together "gauge" a quantity of water which ensures germination and seedling establishment. Even if the seedlings or young plants die before producing new achenes, there is still available a reserve of achenes from previous years.

In a similar way, seed dispersal is delayed in *Anastatica hierochuntica* until there is a heavy shower. After the shower, the dry curved branches enclosing the fruits become straight within two hours (Fig. 122). Following a second shower some of the seeds may be released from the fruits. Skeletons of dead plants of this lignified annual are known to contain viable seeds for many dozens of years (45).

Fig. 21. *Plantago coronopus* — dry infrutescens (left), and 5 minutes after wetting (right).

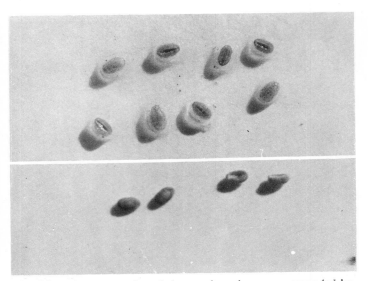

Fig. 22. *Plantago* seeds — below — dry; above — surrounded by mucilage derived from the epidermal cells after wetting.

The gauging of rain is also involved in seed dispersal of *Blepharis ciliaris* (Fig. 130). Each fruit, containing two seeds, is covered by sepals and bracteoles. About 10 minutes after being wetted, the sepals and bracteoles open and the top of the fruit is exposed. The fruit is composed of two halves connected by a tissue layer. After a prolonged rain, when this tissue becomes sufficiently wet, the cell walls in the tissue layer soften. Tension is released in the adjoining fiber tissue causing the softened tissue to tear. The fruit then instantly splits into two halves. During this splitting the dry seeds are projected in opposite directions up to a distance of 5 m from the plant. If the seeds land on wet soil and the temperature is at least 20°C, they may start germinating within two hours. Multicellular hairs which are parallel to the seed surface when dry, become nearly perpendicular when wet by soil or rain. This raises the seed to a position which enables the emerging root to penetrate directly into the soil. Part of this hair tissue becomes mucilaginous after becoming wet. The mucilage helps to anchor the seed to the soil, to keep the area around the germinating seed moist, and to hide the seed from herbivorous animals. The mucilage may also function in long distance dispersal by sticking to bird's feet.

Other hygrochastic plants in the deserts of Israel and Sinai are: *Anvillea garcinii, Cichorium pumilum, Filago contracta* (Compositae), *Mesembryanthemum nodiflorum, M. forsskalii, M. crystallinum, Aizoon hispanicum, A. canariense* (Aizoaceae), *Plantago coronopus* (Figs. 21 and 22), *P. cretica, P. bellardii* (Plantaginaceae), *Salvia viridis, Ziziphora capitata* (Labiatae), *Alyssum damascenum, Notoceras bicorne, Leptaleum fili-folium, Lepidium aucheri* (Cruciferae). *Trigonella stellata, T. monspeliaca* (Papilionaceae).

In addition to the desert hygrochastic species mentioned above, there are ten such species in the non-desert Mediterranean territory of Israel. In the temperate parts of the world, where it rains in summer, hygrochasy, as far as we know, does not exist. The existence of this mechanism in many different families in dry regions is apparently an adaptation to dry environments.

Dispersal of seeds over a long period of time takes place in plants with fruits that do not open when ripe. In such plants the force of raindrops causes some of the fruits to open. During the wet season extensive fungal and microbial activity causes deterioration of the fruit and hence the release of seeds. The fruits of the annual crucifers *Erucaria boveana, Carrichtera annua* and *Torularia torulôsa* remain closed when ripe. Seed dispersal takes place as a result of strong showers.

GERMINATION PROCESS AS WATER GAUGE

The penetration of water into the seed starts the process of germination as long as there are no germination inhibitors in the immediate environment. The water activates the dormant enzyme systems and causes the expansion of certain cells. This activity can be detected when the seeds swell and the embryonic root and shoot emerge.

THE REGULATION OF GERMINATION BY SEED AND FRUIT COATS

Seed coat permeability was studied in two species of Papilionaceae by Gutterman and his co-workers (55, 56, 57). They found that *Ononis sicula* and *Trigonella arabica* produce brown, green, and yellow seeds whose coats differ in structure and in water permeability. Yellow seeds of *T. arabica* have a thicker cuticle and absorb water much more slowly than green seeds. The brown seeds have the thinnest cuticle and swell immediately on contact with water. Day length during the 8—12 final days of seed maturation determines the seed coat characters, so that a single plant may produce all three kinds of seeds. All three types of seeds will germinate if rubbed with sand-paper. This abrasion simulates the activity of fungi and bacteria present in the soil or in the wet fruit. These microorganisms digest away parts of the seed coat. Having seeds of differing permeability in one seed crop extends germination for several years and thus enables the species to overcome periods of unfavorable conditions.

A similar strategy is exhibited by *Hymenocarpos circinnatus*, *Onobrychis squarrosa* and *O. crista-galli*. The diaspore of these plants is a closed fruit containing two or three seeds depending on the species. The seed furthest from the pedicel germinates within the fruit during the first year after ripening; the others germinate in following years only after the seed coat is partially digested by microorganisms. Many species of *Medicago*, *Trigonella* and *Astragalus* also have fruits containing multiseed diaspores in which only one seed germinates in the first year and the remaining seeds during subsequent years.

Many *Atriplex* species have two or more kinds of diaspores which differ in the size and shape of their bracts and in their seed color. It is prominent in *Atriplex dimorphostegia* (77), *A. rosea* (69), *A. leucoclada* and *A. semibaccata*. Experiments have shown that light-colored seeds germinate more rapidly probably because of the higher permeability of their seed coats to water.

THE REGULATION OF GERMINATION BY INHIBITORS

Some desert plants contain water soluble germination inhibitors in their diaspores. Inhibitors have been found in seed coats of *Colutea istria*, in fruit walls of *Zygophyllum dumosum*, *Salsola inermis*, and *S. volkensii* (76, 78), and in bracts of *Atriplex dimorphostegia* (77), *A. halimus*, *A. rosea*, and *Rumex cyprius*. In these plants germination is delayed until there is sufficient moisture to dissolve and leach the inhibitor compounds. Salt is the inhibitor in *Zygophyllum dumosum*. The seed of *Salsola tetrandra* is covered by a thin fruit wall surrounded by 4 sepals and 3 succulent bracts — all of which form the diaspore. The succulent bracts contain salt which remains in them when the seed matures. Only in those few years with sufficient rainfall are the salts leached, allowing seedlings of *S. tetrandra* to develop. However, if a seed is taken out of its diaspore, it germinates within 24 hours.

In several *Aegilops* species the diaspore is a spike containing a few seeds of different sizes. There is a water soluble germination inhibitor in the parts of the spike surrounding the seeds. Each kind of seed in the spike germinates under specific environmental conditions and responds in a different way to the germination inhibitor (32).

HETEROCARPY

The production of morphologically different fruits by the same plant is known as heterocarpy. In heterocarpous species, a single plant produces a variable seed population which may be adapted to different environmental conditions. Heterocarpy is very common in the Compositae, and most species discussed in this section are composites. Each fruit (achene) in the Compositae contains one seed. *Carthamus tenuis* is a summer annual of the transition zone between the desert and Mediterranean ecosystems. It is also a weed in cultivated Mediterranean fields. It has two kinds of achenes. Achenes with a pappus develop in the center of the flowering head and achenes without pappus at the periphery. As the head ripens and becomes dry, the pappus scales of the central achenes spread. As a result, these achenes become detached from the receptacle and are

dispersed by wind. Within a few weeks, achenes without pappus, which are more tightly connected to the receptacle, fall near the mother plant. The pappose achenes are light in color and have a smooth coat; they germinate immediately after ripening. Achenes without pappus are dark and have a rugose coat; they require many days for germination.

Hedypnois rhagadioloides produces three kinds of fruit: 1) a peripheral achene in the logitudinal groove of each involucral bract; 2) in the center of the head thin achenes, each with a small pappus, and 3) between the two other types thick achenes, each with a small rudimentary pappus. After a prolonged rain, the bracts slowly open and the achenes are gradually released. The central achenes which are not attached as securely to the receptacle as the other two types, are released first.

The inflorescence in species of *Calendula* contains several types of achenes. These include achenes with smooth wings, others with serrate margins on the wings, small rounded achenes, and elongated pointed achenes.

Several species of *Crepis*, *Picris*, and *Geropogon* have central pappose achenes easily detached when ripe, and peripherial achenes (with a rudimentary pappus) which are more securely connected to the receptacle. In these species, the peripherial achenes are not discharged until the winter, and fall near the plant.

Amphicarpic plants have two kinds of fruits, *aerial* and *subterranean*. The subterranean fruits mature below the surface of the soil and their seeds later germinate in situ. One such species is *Gymnarrhena micrantha*, an annual composite of the Negev and Sinai. The few subterranean achenes produced by this species are large and germinate readily in a wide range of environmental conditions. The aerial achenes are much more numerous, but have much lower germination rates (80).

The aerial achenes are dispersed in winter following the first shower sufficient to soften the tissue connecting the achenes to the receptacle. In this species the pappus of the aerial achenes irreversibly opens after becoming wet. The achenes are dispersed by wind when dry again. The plants derived from aerial achenes are small and have a low survival rate. The subterranean achenes germinate when the soil surrounding them becomes wet. Soon after the cotyledons reach maximum size, the plants already start blooming. The flower ovary is below the soil-surface whereas the relatively long style, anthers and petals of the small flower reach the surface. During dry years, many of these plants only produce subterranean achenes. Later in the season, if more water becomes available, the plants will continue to grow and produce aerial heads. Populations of *G. micrantha* which develop from subterranean achenes dominate in areas of diffused vegetation on loess plains, marl outcrops, and fine-grained talus slopes. In these habitats the ground is slightly saline. Years of good rainfall result in the production of thousands of aerial achenes. Most of the seedlings derived from these wind dispersed achenes will die. The seedlings which develop in leached soil will give rise to a population which may be temporary because of the competition from other species.

The amphicarpy of *Emex spinosa* (Polygonaceae) was described in detail by Evenari et al. (36). The aerial and subterranean diaspores differ in size, content of germination inhibitors, and sensitivity to light. The two kinds of seedlings also differ in their resistance to drought and their rate of development of aerial diaspores. As in *G. micrantha,* the seedlings derived from subterranean seeds are more viable than seedlings from the aerial seeds. This leads to permanent populations derived from subterranean seeds, and temporary populations from aerial diaspores. The sites with permanent populations are usually depressions in loess plains that receive local flood water.

Amphicarpy is not exclusive to desert plants, and is found in many Mediterranean species of Israel. For a detailed study of the adaptive significance of this mechanism, see Mattatia (90).

Plate 3.   Carob tree *(Ceratonia siliqua)* in a wadi near Mashabbe Sade.

CHAPTER 3. FLORA AND VEGETATION

PLANT GEOGRAPHY

Many attempts have been made to characterize areas in terms of their floral composition. Such areas, known as "phytogeographical regions" or "phytochoria" are delimited by the boundaries of species which occur only in those particular regions. Species which are only present in a given region or in part of it are known as "endemic species" of that region. The fact that certain plants grow in one region and not in others is a result of the influence of past and present environmental conditions.

Three and possibly four phytogeographical regions meet in Israel and Sinai (33, 144). Within Israel, the boundaries of three phytogeographical regions coincide with isohyets. The

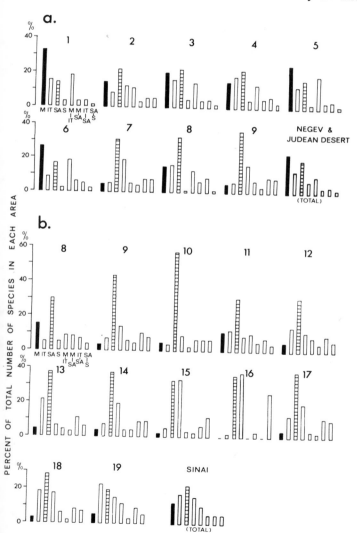

Fig. 23. Phytogeographical analysis of the districts of the Israeli and Sinai deserts. Histograms in part a. represent the districts in Israel, b. apply to the districts in Sinai (numbers correspond to Fig. 7). Abbreviation of the chorotypes: M — Mediterranean; IT — Irano-Turanian; SA — Saharo-Arabian; S — Sudanian; bi-regionals are listed as combinations of the above symbols

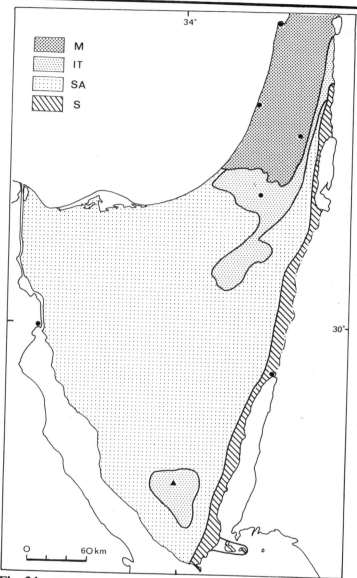

Fig. 24. Plant geographical territories of Israel and Sinai [(the territories in Israel after Eig (33) and Zohary (146)]. Abbreviations as in Fig. 23.

Mediterranean phytogeographical region is bounded in Israel by the 300 mm isohyet. The Irano-Turanian region is situated between the 300 and 80 mm isohyets, while the Saharo-Arabian region dominates areas receiving less than 80 mm of rainfall. Species of the Sudanian chorotype dominate the list of species of Districts 15 and 16 (Fig. 23). Eig (33) and Zohary (44) regarded our districts 2 and 7 as a territory of Sudanian penetration. The boundaries of the entire area, marked as S in Fig. 24, coincide with the isotherm of 23°C.

Each of these phytogeographical regions supports species which also occur in other regions. Such a chorotype is called bi- or pluri-regional.

Figure 23 represents a chorotype analysis of the flora of the 19 districts shown in Fig. 7. The bars in Fig. 23 represent the percentage of the particular chorotype in the total number of species. For example, species of the Mediterranean chorotype (M) account for 33 percent of the total number of species in District 1, whereas the bi-regional Saharo-Arabian and Sudanian chorotype (SA-S) accounts for only 1 percent.

Of the 22 chorotypes in the area under study only the eight most important were included in the chorotype histogram. These chorotypes account for more than 80 percent of total number of species in each district.

The map in Fig. 24, showing the phytochoria of Israel and Sinai, is based on Eig (33), Feinbrun-Dothan (44), and Zohary (144) for Israel, and on original analysis of the flora for Sinai. The chorotypes used in the analysis of the entire area (Fig. 23) are based on Zohary (144, 145) and Gruenberg-Fertig (54). The lists of species in each district are derived from a floristic mapping project by the author.

When comparing the results of our analysis of the whole area (Fig. 23) with Zohary's (144) map (the Israeli part of Fig. 24), we come to the conclusion that the criteria used by Zohary to delimit the phytogeographical regions are not consistent. In delimiting these regions in Israel, Zohary sometimes used floristic composition (the dominant chorotype in the list of species) of the area concerned and sometimes used the extent of ground cover by the dominant species. The use of ground cover parameters to delimit phytogeographical regions by Zohary (143) was criticized by Hedge and Wendelbo (64). As there is no international agreement concerning plant geographical analysis, we present in Fig. 23 the chorotypic composition of each district.

PLANT ASSOCIATIONS

A plant association is defined as "a group of plants with a definite species composition repeated qualitatively and quantitatively" (51). This definition implies that certain species may serve as useful indicators of plant associations. "Exclusive species" occur only in a particular association and thus have a narrow ecological range. Conversely, species that occur in many associations have a wider ecological range. In general, plants that occur in areas having variable amounts of available water can be expected to have a wide ecological range. Therefore habitats having variable unpredictable water regimes will support few, if any, "exclusive species". On the other hand, habitats with a more favorable and constant water regime may support several "exclusive species". Soil pockets of smooth-faced rock outcrops, for example, are steadily supplied with run-in water and are therefore almost independent of changes in annual rainfall. They may be saturated even in years of moderate rainfall and thus are characterized by a rather stable water regime. Such habitats provide suitable conditions for species having a narrow ecological range. Thus, rock vegetation often contains exclusive species.

Plant associations with no exclusive species may be characterized in terms of their dominant species. A species is considered dominant if the surface it covers is greater than that of other species in that area. The dominance of a given species reflects the unique combination of environmental conditions at that location and therefore may be used to divide a heterogeneous area into smaller areas which are ecologically homogeneous. In a given type of environment, an association having a particular dominant species will also include certain other particular companions.

In order to describe an association, records of the vegetation are made in homogeneous sites each having an area of 100 square meters (28, 29). The total proportion of the ground covered by vegetation, and the contribution of the arido-active and arido-passive species to the total cover is listed. The relative contribution of each species to the plant cover is noted as well.

Following the Braun-Blanquet (10) method for classifying vegetation, we cluster together closely related associations into an a l l i a n c e. Allied alliances build up the vegetation o r d e r, with several orders comprising the vegetation c l a s s. The plant communities of the areas with diffuse vegetation fit this hierarchial organization.

The structure of vegetation under extreme desert conditions, such as in Districts 7, 9, 14 and 15, is more complicated. Plant communities replace each other along the wadi in response to changes in water regime. Associations having components belonging to different alliances, orders, and classes may occur in close proximity. Dividing the area into associations according to the method used for diffused vegetation will result in a mixture of small units that cannot be mapped. A suitable method for recording and presenting wadi vegetation was proposed by Lipkin (84). He defined an association as that section of the wadi vegetation typified by the dominance of one or two species. Under extreme desert conditions, each edaphic and physiographic type may be characterized by a typical sequence of associations. The vegetation of the 'Arava Valley was studied by Lipkin's method of sequences (111) and the data will be presented in the following chapter. Complicated sequences are found in areas of hilly topography with diverse types of rocks and slopes. The changing slopes along the wadi lead to changes in water velocity and therefore to changes in water infiltration. The substrate also changes along the wadi. These environmental factors lead to the development of sequences of associations along the wadi. A typical sequence of wadi associations going downstream is as follows:

Annuals occurring only in years of sufficient rain.

Small semishrubs such as *Helianthemum lippii*, *H. kahiricum*, *Fagonia sinaica* or *F. mollis*.

An association of larger semishrubs such as *Artemisia herba-alba*, *Gymnocarpos decander* and *Anabasis articulata*.

Shrubs, such as *Retama raetam*, *Ochradenus baccatus* and *Lycium shawii*.

Trees, such as *Acacia raddiana*, *A. tortilis* or *A. gerrardii*.

Tamarix nilotica close to the base level.

Wadis in gravel plains have more uniform substrates than those in hilly areas. Water velocity is constant and relatively low. The water does not cut into the wadi substrate but rather causes the wadi to widen. Sediments of fine-grained material are deposited. Long stretches of such wadis may have a similar water regime and contain only a few associations in the sequence. In the northwestern part of District 9, in the area southwest of Bir eth Thamada, and in the Paran Plains north-northwest of Elat, wadis varying in size from the 1st order (smallest) to the 4th order support the *Artemisia herba-alba* association. The channel does not become deeper, but divides into several small channels all of which support communities dominated by *A. herba-alba*.

PRESENTATION OF VEGETATION

Because of its diversity, desert vegetation is difficult to depict on small-scale maps. In a general map, such as in Fig. 14, the entire area under discussion is divided into only three units of vegetation — two of diffuse vegetation and one of contracted vegetation. Such a map provides very little information concerning the vegetation. The high diversity of plant communities in many of the districts and the large area covered in this book make it impossible to present accurate vegetation maps. Instead, schematic diagrams are used to illustrate typical patterns of vegetation. Also, a list of associations for each of the geomorphological districts is presented in the following sections.

Plant associations named after their two principal components are written in italics. Associations named after one species (e.g., REAUMURIETUM HIRTELLAE) are written in caps and the suffix — ETUM is added to the genus name of the dominant species. Where two plant communities are listed as inhabiting similar areas, the first-named community is less drought resistant. For wadis having complex or inadequately studied sequences, the dominant plants are listed. The location of each district is indicated by shading in a small map. For further details refer to the original articles (28, 29, 30, 31, 84, 111).

Association	Substrate	Precipitation (mm)
Sarcopoterium spinosum-Astragalus betlehemiticus	limestone	300-400
Sarcopoterium spinosum-Phlomis brachyodon	limestone	300-400
NOAEETUM MUCRONATAE	limestone	100-200
Artemisia herba-alba — Salvia lanigera	limestone	100-150
Anabasis articulata-Halogeton alopecuroides	limestone	70-100
Zygophyllum dumosum-Reaumuria hirtella	limestone	70-100
ONONIDETUM NATRICIS	Senonian chalk and marl	350-450
ARTEMISIETUM HERBAE-ALBAE ONONIDETOSUM	"	300-350
Echinops polyceras-Alkanna strigosa	"	200-300
RESEDETUM MURICATAE	"	150-250
SALSOLETUM VILLOSAE	"	150-250
REAUMURIETUM HIRTELLAE	"	100-200
ATRIPLECETUM GLAUCAE	"	70-100
CHENOLEETUM ARABICAE	"	70-100
SUAEDETUM ASPHALTICAE	"	70-100
SALSOLETUM TETRANDRAE	"	70-100
Salsola vermiculata-Satureja thymbrifolia	calcium silicates	100-150
ANABASETUM SYRIACAE	loess deposits	100-200
Varthemia iphionoides-Origanum dayi	smooth-faced limestone outcrops	100-200
Retama raetam-Rhus tripartita	canyon walls and cliffs	100-200

Along with the increasing aridity from west to east in the Judean Desert, the cover of natural and cultivated vegetation shows a corresponding decrease. North-south belts of vegetation can be detected in this district.

Scattered trees of *Pistacia atlantica, Amygdalus korschinskii, Crataegus aronia* and *Ceratonia siliqua* occur beyond the western boundary of the Judean Desert, at the divide of the Judean Mountains. Such a steppe-forest has been observed at desert margins elsewhere in the Middle East (146).

Associated semishrubs dominate the drier slopes east of the divide where the mean annual rainfall is 250-350 mm. This Mediterranean semishrub vegetation or "batha" (34) covers 30 to 40 percent of the site; it is also known as "phrygana" in other Mediterranean countries. Many batha species penetrate disturbed habitats in the Mediterranean region. *Sarcopoterium spinosum*, for example, is a dominant plant at the western boundary of the Judean Desert and also occurs in large areas in the Judean Mountains, the Carmel and the Galilee. Because this plant cannot survive in the shade of trees, its primary

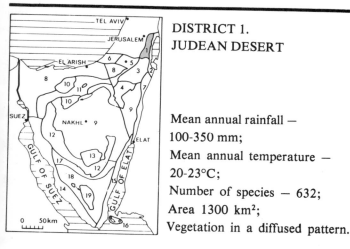

DISTRICT 1. JUDEAN DESERT

Mean annual rainfall — 100-350 mm;

Mean annual temperature — 20-23°C;

Number of species — 632;

Area 1300 km²;

Vegetation in a diffused pattern.

habitat is probably the semi-steppe batha at the margin of treeless desert areas. The Judean Desert batha contains many desert species. *Artemisia herba-alba*, for example, dominates stony soils in areas of 80-150 mm rainfall, and also occurs on chalk outcrops having limited available water. The northern Judean Desert east of the batha belt consists mainly of chalk and marl rocks and the soils derived from these bedrocks. Soils in this area have a salty layer which occurs closer to the surface with increasing aridity. Plant associations in such soils consist of salt resistant semishrubs (halophytes) accompanied by herbaceous plants that are not salt resistant (glycophytes). All of the associations named as occurring on chalk and marl in the above list of associations are dominated by halophytic semishrubs. They are listed in order of increasing resistance to salinity.

Soil salinity is obviously a major factor influencing the occurrence of plant species. The distribution of plant associations reflects the concentration of salts, the depth of the salt layer, and local climatic conditions. However, even in soils that are saline to depths of 20 to 100 cm, the top soil is usually leached and thus may support many glycophytic arido-passive species.

Seedlings of halophytes cannot complete their first year in leached microhabitats because of competition from the glycophytes. Halophytes thus become established only in sites of saline topsoil where the glycophytes are absent. These sites occur near chalk and marl outcrops and in the vicinity of mature halophytic shrubs. Many halophytes absorb and accumulate salts in their leaves and stems. The osmotic gradients so produced enable them to absorb salty water. The fallen leaves and stems of these plants release salts into the topsoil. If more salt is released to the topsoil than is leached by rain, the sites are unsuitable for glycophytes, and halophytes can become established. The salty patches remain for several years following the death of halophytic shrubs. This phenomenon is evident on northern slopes of the Judean Desert where the glycophyte *Poa bulbosa* forms a green carpet except for bare circular or elliptic patches around the halophytic shrubs *Reaumuria hirtella* and *Suaeda asphaltica*.

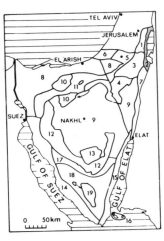

DISTRICT 2.
DEAD SEA VALLEY

Mean annual rainfall —
50—100 mm;
Mean annual temperature —
23°—25°C;
Number of species — 450;
Area — 775 km²
Vegetation in diffused and
contracted patterns

VEGETATION OF THE FAULT ESCARPMENT

Association	Substrate
Zygophyllum dumosum — *Reaumuria hirtella*	hard dolomite or limestone
Gymnocarpos decander — *Iphiona maris-mortui*	blocks of fissured limestone
SALSOLETUM TETRANDRAE	marly conglomerates, marl and stable colluvium
ANABASETUM SETIFERAE	scree of nearly stable colluvium
TRICHODESMETUM BOISSIERII	screes of active colluvium
Anabasis articulata — *Halogeton alopecuroides*	colluvium and marly conglomerates; in gentle slopes — in wadis
SUAEDETUM PALAESTINAE	marly conglomerates, marl and salt marshes
Suaeda fruticosa — *Anabasis setifera*	gravelly shores of the Dead Sea

WET HABITATS

Fresh or brackish water springs:
Belts dominated by *Tamarix nilotica*, *T. tetragyna*, *Phragmites australis*, *Arundo donax* and *Typha australis*.
Salt marshes near the coast and south of the Dead Sea: Belts dominated by *Prosopis farcta*, *Alhagi maurorum*, *Athrocnemum macrostachyum*, *A. fruticosum*, *Seidlitzia rosmarinus*, *Tamarix nilotica*, *Suaeda monoica* and *S. fruticosa*.

WADI VEGETATION

Non-arboreal dominants in the sequences of associations on pebbly ground: *Blepharis ciliaris*, *Trichodesma africana*, *Tricholaena teneriffae*, *Cenchrus ciliaris*, *Salvia aegyptiaca*, *Helianthemum kahiricum*, *Asteriscus graveolens*, *Anabasis articulata*, *Zygophyllum dumosum*, *Ochradenus baccatus*, *Lycium shawii*.

Non-arboreal dominants in the sequences on chalky and marly ground: *Salsola vermiculata*, *S. tetrandra*, *Atriplex halimus*.

Sudanian arboreal components in downstream sections of wadis and near fresh water springs: *Acacia raddiana*, *A. tortilis*, *Ziziphus spina-christi*, *Salvadora persica* and *Moringa peregrina*.

This district contains highly diverse habitats within a relatively small area. The fault cliff escarpments of the Rift Valley comprise many rock types and support diffused vegetation. Constantly falling stones and rocks produce unstable habitats in these escarpments. Some species, such as *Trichodesma boissierii*, *Anabasis setifera*, and *Halogeton alopecuroides*, adapt to this habitat by rapidly becoming established following a disturbance in the landscape. Terraces of the former Lisan Lake support halophytes such as *Salsola tetrandra* and *Suaeda palaestina* that develop in patterns similar to halophyte communities of the Judean Desert. Freshwater springs such as 'Ein Fashkha, which flow out along faults, support hydrophytic species common to similar habitats throughout the world. Such species include: *Phragmites australis, Arundo donax* and *Typha australis*. By comparison, dry habitats of the fault escarpment support *Iphiona maris-mortui*, a species endemic to this district.

The distribution of hydrophytic species is related to salinity level. *Typha australis* is the least salt-tolerant while *Tamarix tetragyna* and *T. nilotica* succeed at sites containing more salt. The dense vegetation at 'Ein Fashkha forms a complex pattern of belts encircling the springs. Species distribution is related here to soil moisture and salinity. The springs near the southern boundary of the district are more saline and true halophytes occur there.

Small fault lines can be recognized from a great distance by the dark color of the halophytes which make use of saline water derived from the fault springs. Impressive stands of date palms occur in small wadis on the white Lisan Marl west and southwest of the Dead Sea Works near Sodom. Their fruits are small and the palms are probably spontaneous here.

'En Gedi is the most interesting oasis formed by a fresh water spring in this district. The springs of 'En Gedi, which flow out high above the level of the Dead Sea, were active when the Lisan Lake existed. These springs probably have supplied fresh water for at least 18,000 years. The area is much warmer than its surroundings, and the combination of high temperature and sufficient moisture enabled many Sudanian species to become established. In addition to *Acacia raddiana, A. tortilis,* and *Ziziphus spina-christi,* which grow in wadis elsewhere in warm parts of the country, the 'En Gedi area also supports thermophytes with high moisture requirements. These include: *Moringa peregrina, Salvadora persica, Cordia sinensis* and *Maerua crassifolia,* all of which have a higher population density here than anywhere else in Israel. These species originated in the African savannas and may have been introduced to the Dead Sea Valley by birds migrating through the rift valley. This explanation is by no means certain, but is supported by the establishment at the Yeroham reservoir in the northern Negev of three species of hydrophytes which normally grow in the Nile Delta. The reservoir was built in 1960. Another hypothesis holds that the Sudanian trees, which prevailed in the Negev during a period with moist and warm climate, continue to survive in the habitats which supply these conditions.

Many plants in small wadis of this district are thermophylous and some have a Sudanian chorotype. The Sudanian component is restricted to habitats where the moisture regime is similar to that of the savannas. Annual vegetation develops in patches on fine-grained alluvium or marl-derived soils in the northern part of the Dead Sea Valley where salt is leached to suitable depths. Saline sites support plant communities of annuals dominated by the following species: *Mesembryanthemum nodiflorum, Salsola inermis, S. volkensii, S. jordanicola, Limonium thounii* and *Aizoon hispanicum*. Each of these is adapted to a particular soil type

and salinity level. *Salsola inermis* dominates on the gravels in the alluvial fan along the Jerusalem — Jericho road at the entrance to the Dead Sea Valley. Another part of the same alluvial fan having silty ground and fewer stones is dominated by *Mesembryanthemum nodiflorum*. The range of salt tolerance of the halophytes is much greater than that of the glycophytes (23). However, in leached soils, the annual halophytes do not compete successfully with glycophytes which use moisture more efficiently.

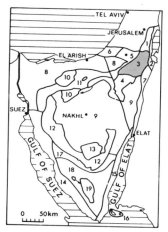

DISTRICT 3. NORTHERN NEGEV HIGHLANDS

Mean annual rainfall — 70—200 mm;

Mean annual temperature — 18°—19°C;

Number of species — 645;

Area — 2175 km²

Diffused vegetation in most of the district

LIMESTONE AND DOLOMITE

Association	Substrate	Precipitation (mm)
Artemisia herba-alba — *Thymelaea hirsuta*	bedded limestone and hard chalk	100-200
Artemisia herba-alba — *Salvia lanigera*	hard chalk	100-150
Artemisia herba-alba — *Gymnocarpos decander*	bedded limestone	100-150
GYMNOCARPETUM DECANDRI	steeply inclined bedded limestone	70-150
Artemisia herba-alba — *Reaumuria negevensis*	bedded limestone with softer interbedded rocks	100-150
Zygophyllum dumosum — *Reaumuria negevensis*	as above, but in south facing slopes	90-100
Zygophyllum dumosum — *Gymnocarpos decander*	bedded limestone or dolomite	90-100
HAMMADETUM SCOPARIAE LEPIDOSUM	Senonian chert covered with stony soil	120-150
Zygophyllum dumosum Reaumuria hirtella	as above, and on bedded and fissured limestone and dolomite	70-150
Zygophyllum dumosum — *Herniaria hemistemon*	calcium silicates	80-100
Anabasis articulata — *Halogeton alopecuroides*	scree in anticline escarpment	70-100

Association	Substrate	Precipitation (mm)
ANABASETUM SETIFERAE	fissured limestone or chert and scree in anticlines escarpment	70-100
Varthemia iphionoides — Origanum dayi	smooth-faced limestone and dolomite outcrops	70-150
Retama raetam — Rhus tripartita	blocky, fissured limestone and dolomite in anti-cline escarpments or canyon walls	70-150

CHALK AND MARL

Association	Substrate	Precipitation (mm)
SALSOLETUM VILLOSAE	chalk of various eras	100-200
REAUMURIETUM NEGEVEN-SIS	mainly Cenoman-Turon chalk	70-100
REAMURIETUM HIRTELLAE	mainly Senonian chalk and marl	70-100
CHENOLEETUM ARABICAE	chalk, phosphatic chalk, calcium silicates	70-100
HALOGETONETUM ALOPECUROIDIS	Senonian chalk and marl	70-100
AELLENIETUM LANCIFOLIAE	alluvium of chalky rocks	70-100
SUAEDETUM ASPHALTICAE	Senonian chalk and marl	70-100
SALSOLETUM TETRANDRAE	in wadis of Nahal Zin with chalky Senonian marl	70-100
HAMMADETUM NEGEVENSIS	in wadis of Nahal Zin with chalky Senonian marl	70-100

LOESS SOILS

Association	Substrate	Precipitation (mm)
ANABASETUM SYRIACAE	uncultivated loess	150-200
HAMMADETUM SYRIACAE	uncultivated loess	80-150
ACHILLEETUM SANTOLINAE	cultivated loess	100-200
Anabasis articulata-Hammada scoparia	uncultivated loess	80-100
Zygophyllum dumosum-Hammada scoparia	uncultivated loess	80-100

SANDY GROUND

Association	Substrate	Precipitation (mm)
NOAEETUM MUCRONATAE ARENARIUM	sandy loess	100-150
ARTEMISIETUM HERBAE-ALBAE ARENARIUM	sandy loess	100-150
ANABASETUM	Neogene sand of	90-150

Association	Substrate	Precipitation (mm)
ARTICULATAE ARENARIUM	Mishor Rotem and Yeroham-Dimona Valley and sandstone outcrops of Makhtesh Hatira	
Six different subassociations distributed in accord with climate and sand depth		
ZYGOPHYLLETUM DUMOSI ARENARIUM	Neogene sandy conglomerate	90-100

WADI VEGETATION

Association	Substrate	Precipitation (mm)
Thymelaea hirsuta-Achillea fragrantissima	2nd to 5th order wadis with silty soil in areas of limestone and loess soil	90-200
Retama raetam-Achillea fragrantissima	similar wadis with pebbly substrate	90-200
Atriplex halimus — Achillea fragrantissima	wadis in areas with chalky ground	
Tamarix nilotica	5th-7th order wadis with pebbly substrate and sites with occasional free water	

This district, with its high diversity of rock types in the anticlines and its varied topography, supports a large number of plant associations, some of which are restricted to small areas. The high diversity of plant communities over a small area may be demonstrated in the eastern flanks of the Rahama anticlinal ridge (Fig. 25). Here five geological formations are exposed along a distance of 300 m.

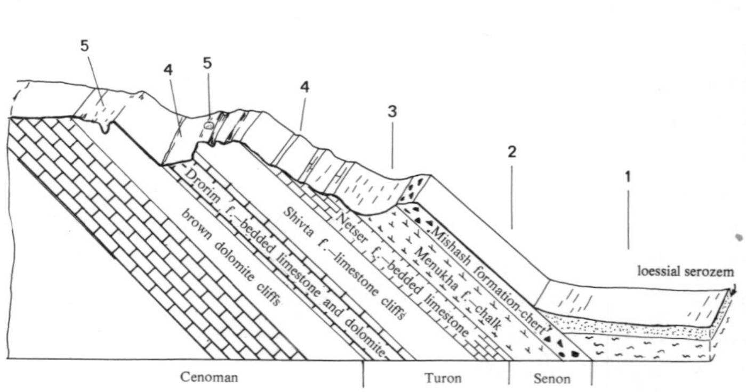

Fig. 25. A substrate-vegetation diagram of the eastern flanks of Rahama Mountain (For explanation, see text).

The following plant communities are identified in Figure 25 according to substrate type:

1. HAMMADETUM SCOPARIAE occurs in the loess plain at the foot of the ridge.

The *Zygophyllum dumosum — Hammada scoparia* association develops in stony-loess and at some sites on the chert slopes of the Mishash Formation.

Plate 4. *Sternbergia clusiana,* a bulbous plant which develops in the Negev Highlands in soil pockets which receive runoff water from smooth-faced outcrops of limestone with a white cover of lichens.

2. GYMNOCARPETUM DECANDRI occupies step-like, highly fissured Senonian chert outcrops.

The *Varthemia iphionoides — Origanum dayi* association occupies poorly fissured chert outcrops. The *Zygophyllum dumosum — Gymnocarpos decander* association occurs where stony soil covers the chert rocks.

3. REAUMURIETUM HIRTELLAE occurs on exposed Senonian chalk. In places where chalk is covered with stony soil, ZYGOPHYLLETUM DUMOSI occupies the southern slopes and ARTEMISIETUM HERBAE-ALBAE occurs on the northern slopes. This is the only habitat in these flanks where slope aspect has a greater influence than rock type on the distribution of vegetation.

4. The *Artemisia herba-alba — Gymnocarpos decander* association is found on outcrops of bedded Turon limestone irrespective of slope aspect. *Thymelaea hirsuta,* which also grows in the wadis of this area, is indicative of the favorable water regime of bedded limestone lacking soil cover. Where soil cover is present, the bedded limestone supports the *Zygophyllum dumosum — Gymnocarpos decander* association.

5. The *Varthemia iphionoides — Origanum dayi* association occurs in the Cenoman and the Turon smooth-faced rock outcrops. Many species comprise this association including *Sternbergia clusiana, Sarcopoterium spinosum, Globularia*

arabica and *Micromeria sinaica. Sarcopoterium spinosum* inhabits wadis with smooth-faced rock outcrops in the catchment area. Neighboring wadis lacking such rocks support the *Thymelaea hirsuta — Achillea fragrantissima* association.

Synclines which were filled with sands, conglomerates, and clays during the Neogene period illustrate other ecological relationships. Here, in the Rotem Plain, vegetation density is higher than in the surrounding hills because of the favorable water regime of the sand . The depth of loose sand above sandstones, conglomerates and clays, influences the water regime and the distribution of vegetation (Fig. 26). Where bare sandstone is exposed, very few arido-active plants occur and the dominant species are *Anastatica hierochuntica* and *Blepharis ciliaris* (30). Moisture is lost either by runoff in the case of heavy rain or through evaporation from the sandstone. In places where the loose sand is up to one meter deep, *Anabasis articulata* is dominant; and *Stipagrostis obtusa, Gymnocarpos decander,* and *Asphodelus microcarpus* are typical associates. In places with a deeper sand cover *Calligonum comosum* is dominant, typically associated with *Stipagrostis plumosa, S. ciliata* and *Echiochilon fruticosum.* In this community a sand mound builds up around each *Calligonum* shrub. These sand mounds indicate that sand movement has taken place in this area.

Hills of conglomerate rocks covered with black chert pebbles support the ZYGOPHYLLETUM DUMOSI ARENARIUM. Psammophytes (sand plants) such as *Neurada procumbens* are among the associates reflecting the sandy matrix of the conglomerate. The dark-colored stones absorb considerable radiation resulting in temperatures high enough for the growth of thermophytes such as *Asteriscus graveolens, Aizoon canariense, Robbairea delileana* and *Commicarpus africanus.* Wadis in this area receive much runoff and support large *Retama raetam* and *Calligonum comosum* shrubs.

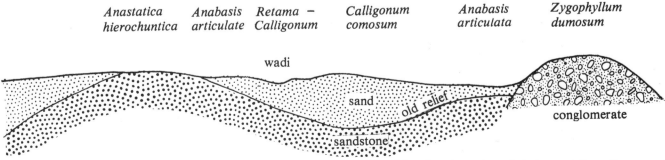

Fig. 26. A schematic substrate-vegetation diagram of Rotem Plain.

Plate 5.   The oasis of 'En 'Avdat near Sede Boqer showing long leaves of *Typha australis* which grows near fresh water. Trees of *Populus euphratica* can be seen in the background.

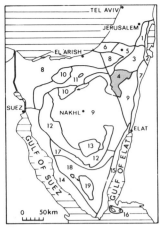

DISTRICT 4. CENTRAL NEGEV HIGHLANDS

Mean annual rainfall —
80-100 mm;
Mean annual temperature —
17°–19°C;
Number of species — 469;
Area — 1775 km²
Vegetation mostly in a diffused pattern.

LIMESTONES

Association	Substrate
Artemisia herba-alba — Reaumuria negevensis	chalky Eocene limestone
Zygophyllum dumosum — Reaumuria negevensis	chalky Eocene limestone in plateaus
Anabasis articulata — Halogeton alopecuroides	bedded limestone at the boundary with District 9
Zygophyllum dumosum — Reaumuria hirtella	bedded Cenoman limestone
Varthemia iphionoides — Pistacia atlantica	smooth-faced hard limestone outcrops

CHALK AND MARL

REAUMURIETUM NEGEVENSIS	Eocene and Turon chalk
REAUMURIETUM HIRTELLAE	Senonian and Meistrichtian marl
CHENOLEETUM ARABICAE	Cenoman and Senonian chalk
SALSOLETUM RIGIDAE	Cenoman-Turon chalk and marl
HALOGETONETUM ALOPECUROIDIS	Chalk and marl in the escarpments of Makhtesh Ramon
HAMMADETUM NEGEVENSIS	Chalk and marl of the northern escarpments of Avdat Plateau
SUAEDETUM VERAE	Chalk outcrops near the springs of 'En Avdat, 'En Aqev and 'En Ziq
SALSOLETUM TETRANDRAE	Chalk and marl

LOESS DEPOSITS

HAMMADETUM SCOPARIAE	elevation below 700 m
ANABASETUM SYRIACAE	elevation above 700 m

WADI VEGETATION

Thymelaea hirsuta — Achillea fragrantissima	2nd to 5th order wadis
Pistacia atlantica-Achillea fragrantissima	3rd to 6th order wadis with gravelly-pebbly ground above 700 m

The most remarkable feature of this district is the occurrence of large *Pistacia atlantica* trees at higher elevations (Fig. 27). Based on counts from aerial photographs, the highest

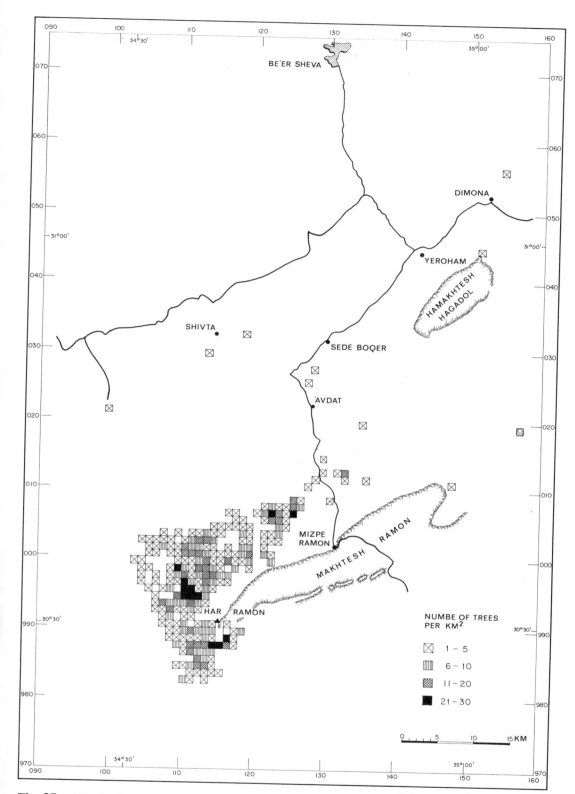

Fig. 27. Distribution map of *Pistacia atlantica* in Negev.

density of trees is found above 700 m. However, elevation is not the most important factor in the distribution of *P. atlantica*. The highest density of trees is associated with large outcrops of smooth-faced limestone resulting in a favorable moisture regime (see pp. 102-104 regarding *P. atlantica)*.

Other trees found in rocks in this district are *Amygdalus ramonensis, Rhamnus dispermus, Rhus tripartita* and *Ceratonia siliqua.* Studies of the pollen assemblages from prehistoric sites (66) reveal that Mediterranean trees such as *Quercus, Olea, Amygdalus* and *Pinus* grew in this district some 10,000 years ago. Climatic conditions have become more arid since then but some Mediterranean relicts have survived in the rock crevices.

Fig. 28. Oblique aerial photograph of Nahal Elot; the line connecting points I, II, III represents the location of the diagram in Fig. 21.

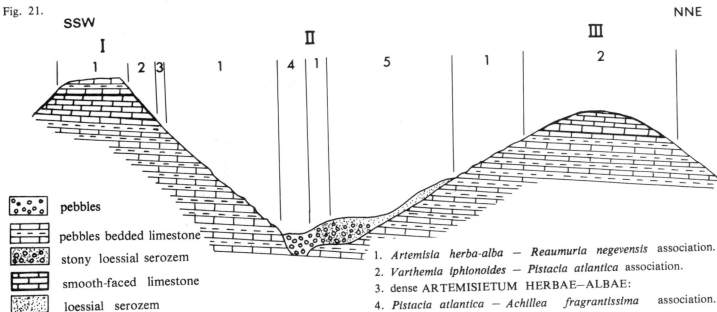

Fig. 29. A schematic substrate-vegetation diagram of Nahal Elot.

pebbles

pebbles bedded limestone

stony loessial serozem

smooth-faced limestone

loessial serozem

1. *Artemisia herba-alba — Reaumuria negevensis* association.
2. *Varthemia iphionoides — Pistacia atlantica* association.
3. dense ARTEMISIETUM HERBAE—ALBAE:
4. *Pistacia atlantica — Achillea fragrantissima* association.
5. ANABASETUM SYRIACAE.

The diagram in Fig. 29 and the photograph in Fig. 28, show the vegetation typical of higher elevations in the district. Slopes of bedded limestone of the Eocene and the Turon (1 in Fig. 29) are populated with the *Artemisia herba-alba — Reaumuria negevensis* association. The most common semishrub companions are *Moricandia nitens, Helianthemum vesicarium* and *H. ventosum*. The most common geophyte and hemicryptophyte associates are: *Rheum palaestinum, Crocus damascenus, Erodium hirtum, Tulipa polychroma, T. amblyophylla, Scorzonera papposa* and *Asphodeline lutea*. In neighboring slopes with chalk outcrops the only semishrub present is *Reaumuria negevensis*.

The smooth-faced outcrops of limestone (2 in Fig. 29) support a plant association that includes many rare plants. Three such plants are *Origanum ramonense, Amygdalus ramonensis* and *Ferula negevensis* which are endemic to this district. Other species growing in these outcrops are probably relics from the more humid climates which once prevailed in the Negev (19, 66). Such species include *Pistacia atlantica, Prasium majus, Astoma seselifolium* and *Sternbergia clusiana,* all of which grow more abundantly in the more humid habitats of the Mediterranean region.

In the Central Negev Highlands, *P. atlantica* trees are found in soil pockets in rock outcrops, at the foot of outcrops, and in wadis which drain rocky slopes. Each of these three habitats has a different water regime which influences the size of the trees growing there. The trees in the soil pockets are smallest, those in the wadis which drain slopes with smooth-faced rock outcrops are the largest and those at the foot of rock outcrops are intermediate in size (Fig. 109).

Loess deposits at the base of slopes support the association dominated by *Anabasis syriaca*. Several plants such as *Ferula biverticellata* and *Leontice leontopetalum* which grow as weeds in Mediterranean fields also grow here. In this habitat they develop on unplowed ground and may be regarded as growing here in their primary habitat.

The spectacular northern escarpments of the 'Avdat Plateau

in the northern part of this district, are famous for the oases in the canyons of Nahal Zin, Nahal 'Aqev, and Nahal Ziq. Springs with fresh to brackish water flow out in these canyons supporting many species not found in any other habitat in the district. Halophytic vegetation occurs at springs where the water has become saline from constant evaporation. At other springs the flow of water is sufficient to prevent salinization. The most important components in the vegetation of the springs are *Phragmites australis, Typha australis, Imperata cylindrica, Agropyrum elongatum, Populus euphratica, Trachomitum venetum, Juncus arabicus, Limonium meyeri, Phoenix dactylifera,* and *Tamarix nilotica*. The last five species withstand more saline conditions than the other species mentioned here.

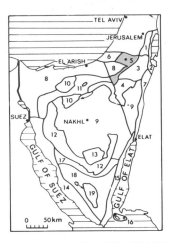

DISTRICT 5.
NEGEV LOWLANDS

Mean annual rainfall —
100-250 mm;
Mean annual temperature —
19°-20°C;
Number of species — 581;
Area — 2025 km²
Vegetation in a diffused pattern

LIMESTONE AND HARD CHALK

Association	Substrate	Precipitation (mm)
NOAEETUM MUCRONATAE	limestone and hard chalk	150-250
Artemisia herba-alba-Thymelaea hirsuta	,,	100-200
Echinops polyceras-Alkanna strigosa	,,	150-250
HAMMADETUM SCOPARIAE LEPIDOSUM	,,	100-150
Zygophyllum dumosum-Reaumuria negevensis	,,	100
REAUMURIETUM NEGEVENSIS	chalk outcrops	100-150

SANDY SUBSTRATE

NOAEETUM MUCRONATAE ARENERIUM	sandy silty soil at the margin of the	150-200
ARTEMISIETUM HERBAE-ALBAE ARENARIUM	sands of District 8	100-150
ANABASETUM ARTICULATAE ARENARIUM		100
PITURANTHOSETUM TORTUOSIS	sandstone outcrops	150-250

Plate 6. *Tulipa polychroma,* a scented flower of the high elevations in the Central Negev Highlands, Gebel Yiallaq and Gebel Halal (northern Sinai) and Gebel el'Igma (central Sinai).

Association	Substrate	Precipitation (mm)
LOESS SOILS		
ANABASETUM SYRIACAE	loess soils	150-200
HAMMADETUM SCOPARIAE	loess soils	100-150
Hammada scoparia-Zygophyllum dumosum	stony loess	100-150
ACHILLEETUM SANTOLINAE (weeds)	cultivated loess	150-250
Kochia indica-Salsola inermis	waste sites of urban areas	

The intensive grazing by Bedouin livestock and the cutting of lignified plants for fuel is reflected in the depauperate appearance of much of the vegetation. In winter many areas appear to be dominated by *Asphodelus microcarpus* (Fig. 6), a plant which is not lignified and not palatable.

Three substrates dominate this district: hard chalk hills, loess soils, and soils mixed with sand derived from District 8. The vegetation of the hills as indicated by remnants of the semishrubs, changes from the relatively wet northern parts where *Noaea mucronata* and *Artemisia herba-alba* are dominant to the drier southern areas where *Zygophyllum dumosum* is dominant. The moist microhabitats of the northern slopes support a Mediterranean batha (as described for the Judean Desert, District 1). Proximity to the Mediterranean territory also increases the incidence of mesic species.

The Negev Lowlands have the most extensive loess deposits in Israel (137), notably in the wide valleys between 'Arad and Be'er Sheva and between Mash'abbe Sade and Nizzana. In the northern portion the soils are cultivated for winter cereals. Many of the wild plants in cultivated areas also occur in non-cultivated loess soils but some are obligatory weeds. *Achillea santolina* and *Hyoscyamus reticulatus* disappear within a few years following cessation of cultivation. However, *A. santolina* also grows in unplowed soils; for example in small wadis and in fine-grained soils in the semi-steppe bathas. This species is probably adapted to a favorable water regime that is promoted in loess soils by disturbance of the soil crust and the increase of soil roughness resulting from cultivation. Phytogeographical analysis of the vegetation of cultivated loess revealed a higher incidence of Mediterranean components than in uncultivated loess (18). The preparation of sites for buildings and the construction of roads expose subsurface layers that often are saline. These disturbed areas are first invaded by xerohalophytes that are dispersed by wind. Some are xenophytic species that become established in the absence of competition from the local plants. Examples include *Kochia indica, K. brevifolia, Atriplex semibaccata* and *Senniella spongiosa*. Other colonizers are halophytes present in other districts but which grow here only in disturbed habitats, e.g. *Salsola inermis, S. volkensii, Atriplex leucoclada* and *Mesembryanthemum nodiflorum*. Within several years the salts are usually leached. Then, the halophytes lose their competitive advantage and the glycophytes take over the area. Disturbed sites that are not saline support colonizers such

as *Lactuca serriola, Conyza bonariensis, C. canadensis, C. albida* and *Aster subulatus* (24) that are wind-dispersed and become established rapidly.

Wind-blown sand from the Haluza Sands (District 8) are mixed with the lithosol developed on hard chalk. The sandy lithosol having a higher water infiltration capacity, has a more favorable water regime than silty lithosol. The dominants of these soils are similar to those of the non-sandy hills, but the shrubs are larger, their density is higher, and the associated species are psammophytes. A component of the NOAEETUM MUCRONATAE ARENARIUM near Nahal Sekher, 15 km south of Be'er Sheva, is *Iris mariae* (Plate 8) which blooms during February and March of rainy years. NOAEETUM MUCRONATAE of the chalk hills north of Be'er Sheva contains the closely related *Iris loessicola* which grows among the *Asphodelus microcarpus* plants.

Trees are rare in this district but occasionally *Acacia raddiana* and *Ziziphus spina-christi*, of Sudanian origin, occur in wadis. In the northwestern part of the district *A. raddiana* grows in a diffused pattern (61).

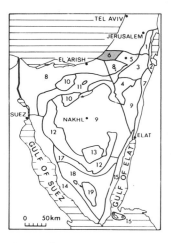

DISTRICT 6.
THE COASTAL PLAIN OF THE NEGEV

Mean annual rainfall — 150-250 mm;
Mean annual temperature — 19°-20°C;
Number of species — 335;
Area — 1200 km²
Vegetation in a diffused pattern.

Association	Substrate
NOAEETUM MUCRONATAE	hard chalk slopes
PITURANTHOSETUM TORTUOSIS	sandstone slopes
LOESS PLAINS	

Most of this area is cultivated and contains weed vegetation. The following winter weeds are found in grain field: *Erucaria boveana, Diplotaxix erucoides, Lolium rigidum* and *Ammi majus*.

The following summer weeds are found in irrigated fields: *Aster subulatus, Conyza bonariensis, Portulaca oleracea, Amaranthus albus, A. retroflexus, A. hybridus, A. palmerii* and *Solanum eleagnifolium*.

SANDY LOESS PLAINS
Artemisia monosperma — Lolium gaudinii
HELIANTHEMETUM SESSILIFLORAE

The weed vegetation here is similar to that of the loess plains but with *Lolium gaudinii — Leopoldia eburnea* as weeds in fields of winter grains.

FOSSIL SOILS IN NAHAL BESOR BADLANDS

Substrate	Association
NOAEETUM MUCRONATAE	fossil loess
SALSOLETUM VILLOSAE,	fossil clay soils
REAUMURIETUM HIRTELLAE	
and ANABASETUM ARTICULATAE	

SPRINGS OF NAHAL BESOR

Dominants near the water include: *Juncus arabicus, Phragmites australis, Typha australis* and *Tamarix nilotica.*

COASTAL DUNES

Stipagrostis scoparia, Convolvulus lanatus and *Artemisia monosperma* are dominants in sands at various stages of stabilization.

Most of this district has been cultivated for many years. Natural vegetation is restricted to the badlands of Nahal Besor and its tributaries and to non-cultivated areas near the boundary of the Gaza Strip. Hence, it is nearly impossible to reconstruct the natural vegetation of the area. A few trees of *Acacia raddiana* and *Ziziphus spina-christi* grow in the area. This suggests that at one time the landscape was dominated by a savanna-like vegetation consisting of annual grasses and scattered trees. The semishrub *Noaea mucronata* may have grown in loess soils as indicated by its present occurrence at undisturbed sites. Closer to the Mediterranean coast, sand content of the soil increases and the dense vegetation is dominated by *Artemisia monosperma* and other psammophytes. Local residents cut *A. monosperma* and use the stems for fuel and for building huts. This species resprouts and germinates rapidly and dense stands occur despite the heavy human pressure.

Between Rafah and Deir el Balah, this vegetation is replaced by cultivated areas and deciduous fruit trees, such as almonds and peaches, grown without irrigation. The sandy loess, several kilometers west of Rafah, is not cultivated and forms the boundary between Districts 6 and 8.

The dense natural and cultivated vegetation of the sandy loess soil contrasts sharply with the lighter sand dunes close to the sea which are bare due to the rough texture of the unstable sand and the high grazing and cutting pressure. The color contrasts are visible from satellite imagery made at an altitude of 920 km (Fig. 8). Plantations of date palms along the Mediterranean coast are also easily seen in Fig. 8.

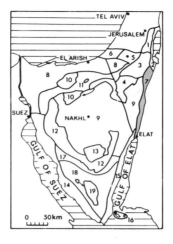

DISTRICT 7. 'ARAVA VALLEY

Mean annual rainfall — 30-50 mm;
Mean annual temperature — 23°-24°C;
Number of species — 350;
Area — 350 km²
Mostly contracted vegetation.

Each major edaphic complex supports a typical sequence of associations along the wadi system as follows:

1. *Anabasis articulata — Acacia tortilis* sequence occurs on well-developed regs with desert pavement of dark chert, in the northern and southern 'Arava Valley.
2. *Anabasis articulata — Acacia raddiana* sequence occurs on well-developed regs at higher elevations of the central 'Arava Valley.
3. *Hammada salicornica — Acacia raddiana* sequence occurs on flood plains of wadis draining the neighboring districts.
4. *Zygophyllum dumosum — Anabasis articulata — Reaumuria hirtella* sequence occurs on Senonian chalk outcrops and alluvial fans of chalk-derived material.
5. *Salsola cyclophylla — Salsola baryosma — Salsola tetrandra* sequence occurs on Lisan Marl badlands in the northern 'Arava Valley.
6. *Hammada salicornica — Calligonum comosum* sequence occurs on slopes of sandy hills.
7. *Haloxylon persicum* sequence occurs on sandy plains, with diffused vegetation in large areas.
8. *Phoenix dactylifera — Juncus arabicus* sequence occurs near springs with fresh or brackish water and the surrounding saline soils.
9. Halophytic vegetation dominated by *Nitraria retusa, Suaeda monoica, Alhagi maurorum* and *Desmostachya bipinnata* grows in belts around the salt marshes.
10. *Tamarix nilotica* and *Tamarix aphylla* grow along the channel of Nahal 'Arava.

The 'Arava contains substantial amounts of subsurface water derived from the large drainage area surrounding it. The abundant water supply is reflected by the dense *Acacia* stands found at many places. Near Hazeva in the northern 'Arava, there is an old large *Ziziphus spina-christi* tree at one of the many springs in the area. This spring seeps through a north-south fault line and south of the tree it supports a belt of halophytic vegetation 20 to 70 m wide and 1,070 m long.

Other springs in the area support large *Acacia* trees. It is evident that direct rain alone does not account for the frequency and size of trees in the 'Arava. The neighboring parts of District 9, for example, which receive similar amounts

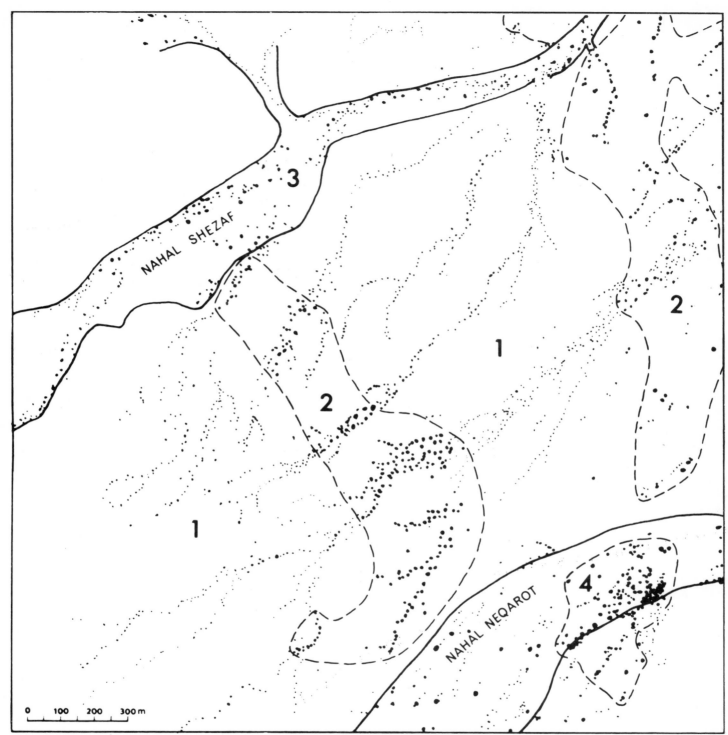

Fig. 30. Tree distribution in a gravel plain near 'En Yahav, drawn from an aerial photograph. 1. Small *Acacia tortilis* trees (2-5 m crown diameter) in wadis of low order. 2. Large A . *tortilis* trees (10-15 m crown diameter) in wadis of low order. 3. Large *A. raddiana* trees in a wadi of high order. 4. A grove of *Tamarix nilotica* in a site with occasional high water table in a large wadi.

of rain, support far fewer trees. When analyzing the vegetation along an ordinary wadi, one comes to a certain section where *Acacia* trees are found. One might expect tree size to increase gradually along the wadi as the amount of available water increases. At many sites, however, there is an abrupt increase in *Acacia* size, as shown in Fig. 30, indicating an underground water supply. Halevy & Orshan (61) found that *Acacia tortilis* is dominant in small wadis and that *A. raddiana* is dominant in large wadis (Fig. 132). In the vicinity of Hazeva all the large trees in some wadis are *A. tortilis* (Fig. 30). This indicates that the trees germinated and became established under the water

regime typical of a small wadi, but grew rapidly when their roots reached underground water. Large *A. tortilis* also occur near Yotvata north of Elat where they benefit from the high water table underlying the alluvium.

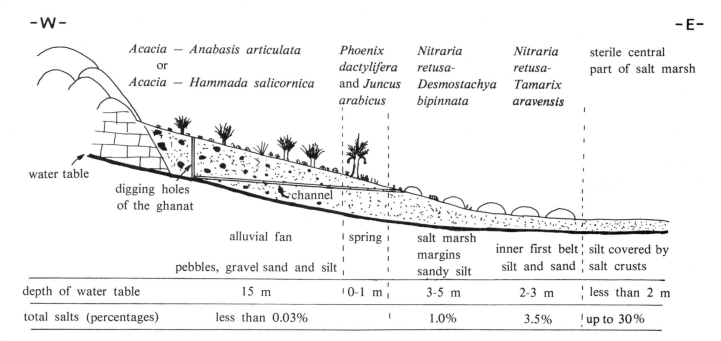

Fig. 31. A schematic diagram of the salt marsh of Yotvata.

The ancient farmers of the 'Arava Valley apparently understood the relationships between large *Acacia* trees, and high water table with low salinity. Remnants of artificial springs ("ghanats" or "foggaras" dug even today in Iran), dating from the Roman and Persian periods, were discovered by Evenari et al. (37) at several sites in the 'Arava. Yotvata was the most developed area in this respect. The ancients dug a well to the water table near the large *Acacia* trees in the alluvial fans and directed the water through a subterranean channel. A series of holes was dug in order to construct a subterranean channel in which the fresh water from the water table could flow to the fields by gravity (Fig. 31). Straight rows of *Desmostachya bipinnata*, a weed of the cultivated plots and the irrigation channel margins, persist to the present day near Yotvata and indicated the layout of the ancient agricultural system. A diagram showing the plant-soil relationship extending eastwards from the western margins of the 'Arava near Yotvata is presented in Fig. 31. Most of the 'Arava consists of gravel plains. Vegetation occurs in the wadis according to the following generalized sequence. At the head of the wadi, in rainy winters the annuals *Aaronsohnia factorovskyi* or *Anastatica hierochuntica* dominate with very few companions. Further down, there is an association dominated by small perennial semishrubs, such as *Helianthemum lippii*, *H. kahiricum*, *Asteriscus graveolens*, or *Blepharis ciliaris*. A lower section is dominated by larger semishrubs such as *Anabasis articulata* or *Gymnocarpos decander* accompanied by many thermophilic species such as *Pulicaria desertorum*, *Tricholaena teneriffae*, *Pergularia tomentosa*, and *Trichodesma africana*. In the lowest sections grow the Sudanian trees *Acacia tortilis*, *A. raddiana*, and occasionally *Ziziphus spina-christi*. This gives the 'Arava the appearance of a savanna, but the trees are restricted to the wadis. The two principal requirements of these trees, high temperature for germination and establishment and sufficient moisture for seedling growth, are satisfied only by wadi habitats.

The three *Acacia* species found in the desert of Israel and Sinai have different environmental demands. *A. tortilis* is the most drought-resistant and the least cold-resistant. *Acacia gerrardii* subsp. *negevensis* is the most cold-resistant, as reflected by its occurrence at high elevations in the Negev and eastern Sinai. *Acacia raddiana* has the widest distribution, growing in moister wadis than *A. tortilis* and in drier wadis than *A. gerrardii*.

Each location within the Afro-Syrian Rift Valley (of which districts 2, 7 and 15 are part) has the highest mean annual temperature for its latitude within Israel and Sinai. In regional temperature maps, isotherms which further south run east-west, here bend northwards. Each species has a northernmost boundary that usually reflects its vulnerability to cold or its requirement for a warm climate (Fig. 32). Of the 16 tree

Acacia raddiana

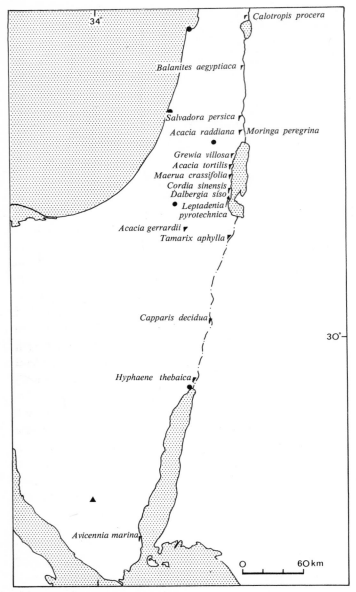

Fig. 32. Northernmost stands of Sudanian or tropical trees in Israel and Sinai.

species given in Fig. 32, ten are found in Israel and Sinai only in the Rift Valley. Their principal habitat is much further south in the East African savannas.

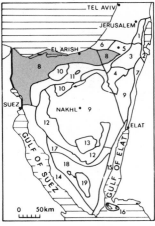

DISTRICT 8.
THE MEDITERRANEAN SANDS AND SALT MARSHES

Mean annual rainfall — 50-150 mm;
Mean annual temperature — 20°C;
Number of species (Negev portion) — 242;
Area (Negev portion) — 650 km²

Number of species (Sinai portion) — 243;
Area (Sinai portion) — 10,050 km²
Vegetation in a diffused pattern.

Association	Substrate
Artemisia monosperma — Thymelaea hirsuta	coarse sand fields with constant removal of fine sand by wind
CORNULACETUM MONACANTHAE	coarse sand fields with constant removal of fine sand by wind
CONVOLVULETUM LANATI	coarse sand fields with constant removal of fine sand by wind
Retama raetam — Astragalus camelorum	stable fine sand fields
STIPAGROSTIDETUM SCOPARIAE	mobile dunes of fine-grained sand on fine-grained
ANABASETUM ARTICULATAE ARENARIUM	alluvium or limestone hills
ZYGOPHYLLETUM ALBI	sand fields on saline soil and outer belts of salt marshes

The halophytic associations of the salt marshes have the following dominants: *Halocnemum strobilaceum, Arthrocnemum macrostachyum, Suaeda aegyptiaca, S. vermiculata* and *Limoniastrum monopetalum; Phragmites australis* and *Juncus arabicus* in brackish parts of the salt marshes.

Two types of substrate predominate in this district — the sands, and the salty soils of the salt marshes. The sandy areas of the dry western portion are depicted schematically in Fig. 33. The finer grains of sand which are constantly removed by

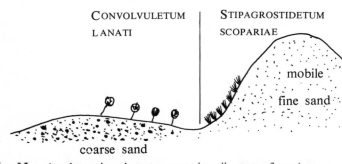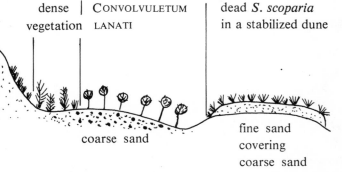

CONVOLVULETUM LANATI   STIPAGROSTIDETUM SCOPARIAE   dense vegetation   CONVOLVULETUM LANATI   dead *S. scoparia* in a stabilized dune

mobile fine sand

coarse sand

coarse sand

fine sand covering coarse sand

Fig. 33. A schematic substrate-vegetation diagram of sandy area near Tasa, northwestern Sinai.

Fig. 34. *Convolvulus lanatus* a plant of sandy areas which withstands the exposure of its taproot.

wind accumulate in dunes. These dunes move in the direction of the prevailing winds. In the coarse sand fields where there is constant removal of sand by wind, the dominant plants are semishrubs possessing taproots capable of resisting desiccation after being exposed. The taproots of *Convolvulus lanatus* (Fig. 34) and *Artemisia monosperma* each have a thick corky bark which protects the vascular tissues. In roots of *Cornulaca monacantha,* as in other Chenopodiaceae, xylem and phloem elements are not restricted to the periphery but are distributed throughout the root. This allows the plant to survive even when the roots become dry at the periphery. These adaptations for resisting desiccation allow up to 2 m of the taproots of these shrubs to be exposed.

The dominant plant of the mobile sand dunes and sometimes the only species present is the perennial grass *Stipagrostis scoparia.* It grows only in sites that are continuously being covered by new sand and dies if the sand cover is removed by wind or if sand coverage ceases. Tops of dunes are bare of vegetation probably because of the highly mobile sand. The most vigorous vegetation and the largest number of mesophytes in the area are found near the base of dunes. These habitats are protected from wind erosion and have a favorable water regime that includes water not captured by the sparse

vegetation of the dune. In years with sufficient rain, the Bedouin plant extensive areas of dunes and sand fields, with watermelon.

A rapid, artificial stabilization of sands occurred at the eastern margin of the district near Nahal Sekher, 18 km south of Be'er Sheva.

During 1965-1966, *Tamarix aphylla* trees were planted west of a new road, to protect it from drifting sand. The resulting decrease in wind velocity caused a considerable change in vegetation and soil east of the road. A reduction in sand accretion improved the germination of *Artemisia monosperma* seeds which require only a thin cover of sand sufficient to prevent desiccation but not so thick as to prevent light penetration (81). *Stipagrostis scoparia,* the former dominant of the area which previously has the highest sand mobility (Fig. 35), gave way to *A. monosperma.* Because of the short life span of *Stipagrostis scoparia* roots, the stems have to be covered with sand in order to produce new roots. As a result of the reduction in wind velocity, there was a greater deposition of silt and clay from dust storms, thus improving surface moisture conditions (26). A biological crust developed, composed of blue-green algae, fungi, and mosses (Fig. 36). This crust binds the sand and decreases its mobility. The increased vegetation further reduced wind velocity near the ground and accelerated the deposition of additional fine-grained material. The higher proportion of silt and clay results in increased biomass production and species diversity. Those sand fields, which were stable long before the road was constructed support the *A. monosperma — Retama raetam* association accompanied by dense stands of the geophyte *Iris mariae.*

Fig. 35. A dune with mounds of sand accumulated around *Stipagrostis scoparia.* Note the ripples which indicate sand mobility (30 km south of Beér Sheva).

Fig. 36. A sand dune 10 years after stabilization began. The mounds was accumulated around a tuft of *Stipagrostis scoparia* which gave way to *Artemisia monosperma*. Note the dark-colored biological crust on the slopes of the mounds.

Overgrazing and cutting of the sand vegetation by Bedouin, west of the June 1967 cease-fire line, led to development of a rather light-colored area in Sinai (Fig. 8). The vegetation east of this line was not cut and therefore the area looks dark. Otterman et al. (104) found that these differences lead to more rain over the vegetated dark area. The vegetation on parts of this district was described by Orshan and Zohary (103).

Plate 7. A salt marsh west of El'Arish with white salt in the center, a brown zone where *Juncus arabicus* is dominant, a green zone of *Halocnemum strobilaceum,* and date palms planted at the margins of the last vegetation zone.

Halophytic vegetation dominates near the Mediterranean coast and at salt marshes up to 10 km south of the coast. Seasonal fluctuations in the depth of the water table and in water salinity greatly influence the distribution of vegetation. A typical salt marsh west of El 'Arish (Plate 7) has a belt of ZYGOPHYLLETUM ALBI on sands near the salt marsh and planted date palms. These palms depend on fresh or brackish water at the root zone. An inner belt is dominated by *Halocnemum strobilaceum.* A zone near the white center contains *Juncus arabicus* plants, most of which are dead probably as a result of a recent increase in salinity. This species grows at sites of moderate salinity.

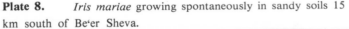

Plate 8. *Iris mariae* growing spontaneously in sandy soils 15 km south of Be'er Sheva.

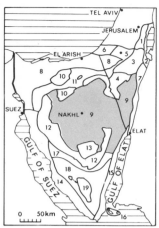

DISTRICT 9.
GRAVELLY PLAINS OF CENTRAL SINAI AND SOUTHERN NEGEV

Mean annual rainfall—25-50 mm;
Mean annual temperature — 18°-20°C;
Number of species (Negev portion) — 227;
Area (Negev portion) — 3,050 km²
Number of species (Sinai portion) — 267;
Area (Sinai portion) — 17,570 km²
Vegetation in a contracted pattern.

The dominant trees in the lower parts of the wadis are: *Tamarix nilotica*, *T. aphylla*, *Acacia gerrardii* subsp. *negevensis* and *A. raddiana*.

Because of the level topography, the distribution of vegetation is best seen in aerial photographs or LANDSAT imageries. From this aerial view various colors can be distinguished, each color corresponding to a particular substrate. The Eocene chalk north of Gebel el 'Igma, containing interbedded chert layers, weathers to reg soils covered with black chert gravel (Fig. 37). Soft chalk and marl, each without chert, are much lighter in color and contain wadis with deeper channels.

Each substrate also has a characteristic wadi pattern (Figures 38, 39, 40, and 41) and usually supports a distinctive vegetation. A wadi may change in width along its course, generally accompanied by changes in vegetation. Figures 40 and 41 demonstrate some general features of wadis in extremely arid areas.

Fig. 37. Chert gravel covering fine-grained soil in a reg in central Sinai. a. general view. b. a closer view of a.

The wadi vegetation is dominated by the following semishrubs and shrubs: *Anabasis articulata*, *Artemisia herba-alba*, *Pituranthos tortuosus*, *Zygophyllum dumosum*, *Z. coccineum*, *Z. album*, *Hammada scoparia*, *Gymnocarpos decander*, *Zilla spinosa*, *Salsola tetrandra*, *Anabasis setifera*, *Reaumuria hirtella*, *Retama raetam* and *Lycium shawii*.

Fig. 38 Aerial view of a wadi system in Central Sinai, a typical branching pattern. Note the dark color of the chert.

Fig. 39. Wadi system in light-colored chalk and marl, at the foot of Gebel Budhiya, showing dense branching and deep channels.

Fig. 40 Aerial view of a wadi system showing the splitting and converging of wadis.

In wadis with deep channels, vegetation is generally found only on the banks, and is often dominated by *Tamarix nilotica* trees (Section 1 in Figures 40 and 41). Trees at the center of the water channel are subject to considerable erosion and are carried away by floods when their roots are exposed. The water channel of a wadi may split up into many small channels, spreading the water over a larger surface in the wadi bed. Such areas are dominated by semishrubs. On a soft substrate the water will tend to dig a deep channel (Fig. 39), while on a hard substrate (such as the compact alluvium in Fig. 42 or Fig. 40, Section 3), the water flows in smaller shallow channels. At Section 3 (Figs. 40 and 41) many small channels converge and support semishrubs. Section 4 denotes an old flood plain or wadi terrace which is elevated and not subject to flooding. This area supports very few plants. The darker color of this habitat is due to in situ weathering of water-transported pebbles.

The large wadis which drain the entire district include Wadi el 'Arish, Wadi Quraiya and Wadi el Bruk. Most floods in these wadis flow in the main channel which looks like Section 1 in Fig. 41. Occasional heavy floods overflow and cover the 1 to 5 km wide flood plain which looks like Sections 2 and 3 in Fig. 41.

Some wadis are flat enough so that the flood water spreads fairly evenly. These sites are cultivated by Bedouin. They are sown with barley and wheat in the event of early winter floods, and with sunflower and sorghum when the floods occur in late winter.

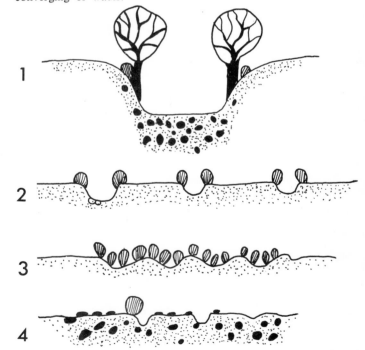

Fig. 41. Schematic diagrams of sections of the wadi system in Fig. 40 .

1. A deep channel where tamarisk trees are dominant. 2. Several branches resulting from the splitting of the channel in 1 supporting semishrubs. 3. Flood plain dissected by many channels close together. 4. An old flood plain now almost devoid of vegetation.

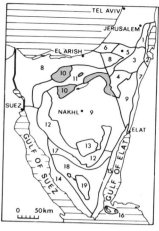

DISTRICT 10.
TRANSITION BETWEEN
DISTRICTS 8 AND 9

Mean annual rainfall —
50-100 mm;
Mean annual temperature —
18°-21°C;
Number of species — 93;
Area — 4,700 km²
Vegetation pattern varies
according to edaphic conditions.

Fig. 42. Wadi system showing water flowing in small shallow channels a few minutes after a strong shower.

Association	Substrate
Anabasis articulata — *Artemisia monosperma*	wide sandy wadis
ANABASETUM ARTICULATAE ARENARIUM or	sand covering alluvium or limestone hills
RETAMETUM RAETAMI	
PANICETUM TURGIDI	sand fields or wadis with sandy ground
ZYGOPHYLLETUM ALBI	sand covering salty ground
STIPAGROSTIDETUM SCOPARIAE	mobile sand dunes

Wadi vegetation dominated by: *Anabasis articulata, Fagonia arabica, Farsetia aegyptiaca, Artemisia herba-alba* and *Artemisia monosperma*.

The vegetation of this district has not been studied in great detail and the number of species may be double that recorded (cf. Fig. 61). This district is strongly influenced by sand blowing inland from the coast, and/or mixing with the existing substrates (Fig. 43). The western, sand-covered hills support stands of large *Retama raetam* shrubs. This plant usually grows in wadis, but develops on these hills because of the high infiltration capacity of the sand together with the favorable moisture regime of the limestone substrate. *R. raetam* shrubs may be covered with up to 2 m of sand, or completely exposed, depending on wind direction. There are usually no other plants accompanying the *R. raetam* here.

prevailing winds
→

Hilltops and other sites having less than 70 cm of sand cover support a plant community dominated by *Anabasis articulata*. Only perennial plants grow in this habitat because of the constantly shifting sand. Where the sand cover is at least 2 to 3 m, there is typical sand vegetation, with *Stipagrostis scoparia* dominating the mobile sand dunes. The seeds of this plant are dispersed by wind and accumulate on the leeward side of dunes (Fig. 43). The seeds are covered by fresh sands and germinate en masse after a rain.

Sand fields near Gebel Libni support a community dominated by *Panicum turgidum*, an important perennial grass of sandy soil in the 'Arava Valley and southern Sinai. This grass also dominates in sand-covered wadi beds. In a similar area, west of the Suez Canal (21), the *Panicum turgidum* plants which grow in the wadi trap blowing sand. This sand improves the water regime locally by increasing water infiltration.

Large wadis with sandy-silt support the *Anabasis articulata — Artemisia monosperma* association. In large wadis with good distribution of water throughout the width of the wadi, the Bedouin may cut the shrubs and grow crops. The district also contains sandless gravel plains that support vegetation similar to District 9.

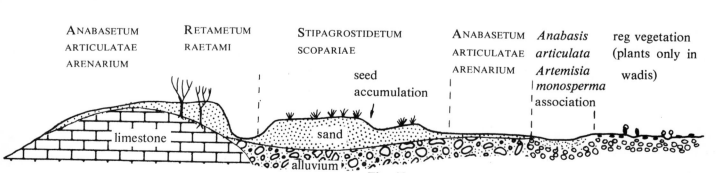

| ANABASETUM ARTICULATAE ARENARIUM | RETAMETUM RAETAMI | STIPAGROSTIDETUM SCOPARIAE | ANABASETUM ARTICULATAE ARENARIUM | *Anabasis articulata Artemisia monosperma* association | reg vegetation (plants only in wadis) |

seed accumulation

limestone sand alluvium

Fig. 43. A schematic presentation of the prevailing habitats of District 10, and their vegetation.

Fig. 44. *Retama raetam* growing in a diffused pattern on limestone hills frequently covered with sand (northwestern Sinai).

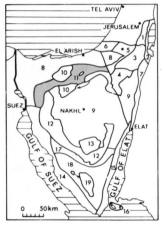

DISTRICT 11. ANTICLINES OF NORTHERN SINAI

Mean annual rainfall — 50-100 mm;
Mean annual temperature — 16°-20°C;
Number of species — 374;
Area — 3,000 km²
Vegetation mostly in a diffused pattern.

LIMESTONE AND CHALK

Association	Substrate
Zygophyllum dumosum — Reaumuria hirtella	bedded limestone
Zygophyllum dumosum — Reaumuria negevensis	bedded limestone
Zygophyllum dumosum — Artemisia herba-alba	bedded limestone
GYMNOCARPETUM DECANDRI	bedded limestone
Artemisia herba-alba — Noaea mucronata	bedded limestone
Anabasis articulata — Halogeton alopecuroides	bedded limestone
STACHYDETUM AEGYPTIACAE	smooth-faced outcrops of limestone
IPHIONETUM MUCRONATAE	smooth-faced outcrops of limestone
GLOBULARIETUM ARABICAE	smooth-faced outcrops of limestone
VARTHEMIETUM IPHIONOIDIS	smooth-faced outcrops of limestone

Association	Substrate
Juniperus phoenicea — Varthemia iphionoides	smooth-faced outcrops of limestone
SUAEDETUM PALAESTINAE	chalk and marl outcrops
SUAEDETUM VERAE	chalk and marl outcrops
REAUMURIETUM HIRTELLAE	chalk and marl outcrops
REAUMURIETUM NEGEVENSIS	chalk and marl outcrops
CHENOLEETUM ARABICAE	chalk and marl outcrops
SALSOLETUM SCHWEINFURTHII	chalk and marl outcrops
SALSOLETUM TETRANDRAE	chalk and marl outcrops

SANDY GROUND

HALOXYLETUM PERSICI	Lower Cretaceous sands of Gebel Maghara
HAMMADETUM SALICORNICAE	as above
Artemisia monosperma — Fagonia arabica	sandy-silt
ANABASETUM ARTICULATAE ARENARIUM	sand on limestone hills or silty alluvium

WADI VEGETATION

Retama raetam — Achillea fragrantissima in large wadis (4th and 5th order) in limestone areas. *Acacia raddiana, Acacia gerrardii* ssp. *negevensis* and *Tamarix nilotica* or *Tamarix aphylla* grow in large wadis.

A variety of substrates and habitats are found in these anticlines. They are somewhat similar to anticlines in the northern Negev (District 3), but the climate is drier here and therefore the plant communities are different. The vegetation of Gebel Maghara was described by Shmida & Orshan (119).

A diagram showing the substrate-vegetation relationship at the Gebel Halal anticline, is given in Fig. 45. The easiest approach to G. Halal is by way of the Quseima-Umm Qataf road parallel to the southeastern flanks of the anticline. The unpaved road leads to an erosion crater. Some million years ago here was the highest peak of the anticline. The old strata at the bottom of the crater are covered with alluvium derived from the weathering of adjacent ridges (Fig. 45, Section 1). The crater floor supports vegetation in its wadis. The *Anabasis setifera* association is found at the upper smaller parts of the wadi and is replaced by the *Retama raetam — Achillea fragrantissima* association along the lower larger sections of the wadi system. Areas in the crater floor used for Bedouin encampments are dominated by *Anabasis syriaca*. This is a xerohalophyte that can tolerate high concentrations of nitrogen.

The slopes inside the crater (Section 2 in Fig. 45) are covered with stony alluvium which, because of the favorable water regime, supports diffused semishrub vegetation. The southern slopes of some alluvial hills support bunches of *Caralluma sinaica,* a stem-succulent plant that looks like a cactus but belongs to the Asclepiadaceae (cf. Fig. 18). Most of the semishrubs on the slopes are xerohalophytes, such as *Reaumuria hirtella, Anabasis setifera, Halogeton alopecuroides, Atriplex leucoclada,* and *Suaeda palaestina.*

-NW-   -SE-

number of section	4	3	2	1
Habitat	Wadi Abu Seyal shallow to deep wadis with gravels	slopes of bedded limestone, chalk and marl, cliffs and smooth-faced outcrops of limestone and dolomite.	slopes of alluvium and rock debris in the crater escarpments	crater floor, alluvium and gravel plains
Vegetation	*Acacia gerrardii, Anabasis articulata Zygophyllum dumosum, Thymelaea hirsuta*	semishrubs covering the slopes in a diffused pattern include: *Zygophyllum dumosum, Artemisia herba-alba, Anabasis articulata, Suaeda palaestina.*	*Halogeton alopecuroides*	ANABASETUM SETIFERAE in small wadis, *Retama raetam-Achillea fragrantissima* in large wadis, *Anabasis syriaca* near Bedouin encampments.

m

Legend:

alluvium

limestone

sandstone

chalk and marl

Fig. 45. A schematic substrate-vegetation diagram of Gebel Halal, northern Sinai.

Fig. 46. *Juniperus phoenicea* growing in: 1. Crevices of smooth-faced limestone. 2. at the foot of the outcrop and 3. in a small wadi receiving water from the rocky slope.

In the northwestern flanks of Gebel Halal (Section 3 in Fig. 45), bedded limestone, chalk, marl, hard limestone, and hard dolomite are exposed. Wadis cutting through these flanks produce slopes facing various directions. Limestone on south-facing slopes, and marl and chalk on all slopes support mixed or monospecific associations of the xerohalophytic semishrubs: *Reaumuria negevensis, R. hirtella, Salsola schweinfurthii, S. tetrandra, Halogeton alopecuroides, Atriplex glauca,* and *Suaeda palestina.* The actual species which occur are determined by soil depth, type of bedrock and slope angle. The bedded limestone on north-facing slopes above 800 m are mostly populated with the more mesophytic association of *Artemisia herba-alba* and *Noaea mucronata.* In the spring of rainy years this association is accompanied by the geophytes *Tulipa polychroma* (Plate 6), *Anemone coronaria,* and *Ranunculus asiaticus,* and by many annuals.

The most mesophytic flora on these northwestern flanks is found on smooth-faced outcrops of hard limestone and dolomite. *Juniperus phoenicea,* a Mediterranean tree, grows in crevices of these outcrops (Fig. 46) as well as in wadis (Fig. 47). Some of the *Juniperus* trees in wadis attain a height of 10 to 12 m, and individuals 4-8 m high are common in all the wadis (Fig. 47). Vines of *Ephedra aphylla* cover many trees, giving the vegetation a Mediterranean appearance. (Additional details on the juniper are given on page 104). An especially interesting plant accompanying *J. phoenicea* in rock habitats is *Origanum isthmicum* (Fig. 48). This species is endemic to the western flanks of Gebel Halal (17). The entire world population of 1,000 to 2,000 individuals occurs within an area approximately five by two kilometers on the northwestern flanks of G. Halal. Mediterranean relicts are also associated with the junipers. These include *Ephedra campylopoda, Astoma seselifolium, Rubia tenuifolia, Sternbergia clusiana,* and *Pancratium parviflorum* (19). The lowest parts of the northwestern flanks largely support associations of *Zygophyllum dumosum.*

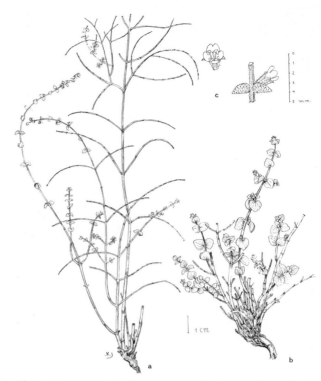

Fig. 48. *Origanum isthmicum,* an endemic species of Gebel Halal.

At the foothills of Gebel Halal there is a sharp transition from the alluvial fans to the sandy area of District 10 (Fig. 45, Section 4). Alluvium in the wadis support *Acacia gerrardii* trees, which do not grow in the sands. This species is even more common in the area around Kuntilla, 80 kilometers southeast of here. Other anticlines support the more common *Acacia raddiana.* The distribution of these two species does not reflect the present pattern of seed dispersal, but may have come about when drainage systems of the Tertiary period carried seeds to the present tree locations (59).

Fig. 47. A large tree of *Juniperus phoenicea* in a wadi of 5th order.

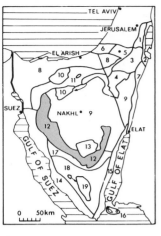

DISTRICT 12.
TABLE MOUNTAINS OF WESTERN AND CENTRAL SINAI

Mean annual rainfall —
50-100 mm;
Mean annual temperature —
16°-20°C;
Number of species — 361;
Area — 6,000 Km²
Vegetation both diffused and contracted.

DIFFUSED VEGETATION

Association	Substrate
Zygophyllum dumosum — Reaumuria hirtella	bedded limestone
Zygophyllum dumosum — Reaumuria negevensis	bedded limestone
Zygophyllum dumosum — Salsola cyclophylla	bedded limestone
Zygophyllum dumosum — Gymnocarpos decander	bedded limestone
ANABASETUM ARTICULATAE	bedded limestone
Artemisia herba-alba — Gymnocarpos decander	bedded limestone
Artemisia herba-alba — Hammada scoparia	bbedded limestone
Artemisia herba-alba — Noaea mucronata	bedded limestone
HAMMADETUM SCOPARIAE LEPIDOSUM	bedded limestone
ANABASETUM SETIFERAE and HALOGETONETUM ALOPECUROIDIS	colluvium in escarpments
Varthemia iphionoides — Gymnocarpos decander	smooth-faced outcrops of limestone

Association	Substrate
Varthemia iphionoides — Stachys aegyptiaca	chalk and marl
ATRIPLICETUM GLAUCAE	chalk and marl
CHENOLEETUM ARABICAE	chalk and marl
Halogeton alopecuroides — Salsola schweinfurthii	chalk and marl
REAUMURIETUM HIRTELLAE	chalk and marl
REAUMURIETUM NEGEVENSIS	chalk and marl
SALSOLETUM TETRANDRAE	chalk and marl

CONTRACTED VEGETATION

The smaller (1st to 4th order) wadis are dominated by: *Artemisia herba-alba, Hammada scoparia, Zygophyllum dumosum, Gymnocarpos decander, Anabasis articulata, Salsola tetrandra.*

The larger (5th to 6th order) wadis are dominated by: *Retama raetam — Achillea fragrantissima* association.

Most of this hilly district consists of plateaux, but dome and anticlinal structures with inclined rock strata are also present. The inclination of the strata influences weathering patterns, water regime and vegetation. A substrate-vegetation diagram of Gebel et Tih (Fig. 49) shows that the drought-resistant *Zygophyllum dumosum* association is replaced by the *Artemisia herba-alba* association at higher elevations. The vegetation of these high elevations is diffused, as would be expected from the many days of high relative humidity resulting in mesic conditions. During a visit to the site in the summer of 1967, dense clouds were observed at the summit for several hours following sunrise. The rocks covered with crustose lichens were wet. Kappen and Lange (70) found that desert lichens obtain most of their moisture from fog and dew.

Hammada scoparia is dominant on marl outcrops having a salt regime similar to that described in District 1. *H. scoparia* also dominates the saline loess soils of the Negev Highlands in terraces cultivated hundreds of years ago. Since *H. scoparia* is also dominant and widespread in saline soils that were never

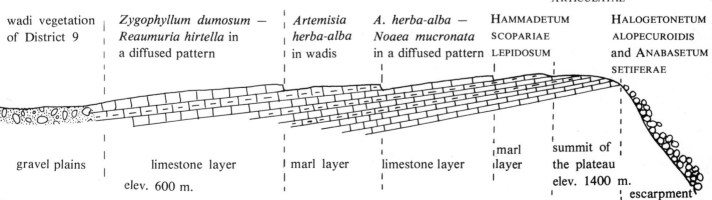

Fig. 49. A schematic vegetation-substrate diagram of Gebel et Tih at the longitude of 33°30'E.

Fig. 50. A spring in Wadi el Shalalah with date palm and *Juncus arabicus* tufts.

cultivated, it may be regarded as a xerohalophyte rather than a weed.

The steep escarpments of the Tih plateau support pioneer associations dominated by *Halogeton alopecuroides* or *Anabasis setifera*. Plants of these species produce many seeds that are dispersed by wind, and colonize the new ground formed by landslides and extensive erosion in the steep slopes. These species also dominate some saline habitats not in the escarpments.

The vegetation of the plateaux varies with the type of rock. *Artemisia herba-alba* or *Zygophyllum dumosum* dominates those slopes consisting of hard rock containing little marl. Small plateaux of horizontally bedded hard strata (such as Gebel er Raha, Gebel Budhiya and Gebel Sahaba) are dominated by the HAMMADETUM SCOPARIAE LEPIDOSUM. Softer rocks support *Zygophyllum dumosum* along with *Salsola cyclophylla* and other xerohalophytes. This is typical of the western part of Gebel er Raha near District 14. Western Gebel Budhiya, which consists of softer rocks supports the *Halogeton — Salsola schweinfurthii* association on its slopes, and HAMMADETUM SCOPARIAE in wadis at the summit of the plateau.

Outcrops of smooth-faced limestones are restricted to the dip-slopes of inclined rock strata at Gebel el Giddi and near the Mitla Pass. *Haplophyllum poorei* which occurs here, is a rare plant confined to rocks, and has previously been found in the Central Negev Highlands (District 4) and the Edom Mountains in Transjordan. It also grows on inclined rocks in Gebel Sahaba and Gebel Heitan in this district. *Pistacia atlantica* trees (cf. Fig. 107) occur in wadis of Gebel Sahaba and Gebel Shaira. One *Pistacia* tree at Gebel Sahaba is the only Sinai tree known to contain *Loranthus acaciae,* a very common parasite of *Acacia* in the 'Arava and Dead Sea Valleys.

Springs, a common feature of this district, result from the alternation of limestone and marl layers and from the many faults. Dozens of springs occur throughout the western part of the district, and others are found in canyons (e.g., Moyet el Gulat near Ras el Gindi, and 'Ein Abu Ntegina). Other springs like those, in Wadi Shalala (Fig. 50), are found in wadis of flatter areas without canyons. Most springs can be recognized by the presence of date palms, many of which produce small fruits with little pulp. Other trees have large fruit typical of palms cultivated in the area of 'El Arish. The suggestion that such springs are a primary habitat of the date palm is discussed later (pp. 119-120).

Fig. 51. The escarpments of Gebel el 'Igma.

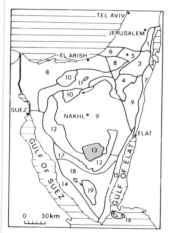

DISTRICT 13.
GEBEL EL 'IGMA

Mean annual rainfall — 50 mm;
Mean annual temperature —
15°-18°C;
Number of species — 186;
Area — 1300 km²
Contracted vegetation at an
elevation of 600-1100 m, and
diffused vegetation at an
elevation of 1200-1600 m.

CONTRACTED VEGETATION

The wadi vegetation is dominated by the following
semishrubs: *Salsola tetrandra, Halogeton alopecuroides,
Reaumuria hirtella, Artemisia herba-alba, Hammada
salicornica, Eurotia ceratioides.*

The larger (4th to 6th order) wadis are dominated by:
Atriplex halimus — Achillea fragrantissima association;
Retama raetam — Zilla spinosa association.

DIFFUSED VEGETATION

The diffused vegetation consists of:

SALSOLETUM TETRANDRAE
HALOGETONETUM ALOPECUROIDIS
SALSOLETUM RIGIDAE
ANABASETUM SYRIACAE

Chenolea arabica — Atriplex glauca association
Artemisia herba-alba — Atriplex leucoclada association
Salsola tetrandra — Hammada negevensis association

Gebel el 'Igma is a chalk plateau with a gradual north to
south ascent from 600 m in the north to 1,600 m in the south.
The vegetation changes gradually with elevation and most of
the dominants are xerohalophytes. Glycophytes, such as
Artemisia herba-alba and *Hammada salicornica*, occur in old
terraces of the main wadi. *A. herba-alba* dominates gravelly
terraces where the chalk material was leached and the
substrate has a water and salt regime similar to the regs of
central Sinai. *H. salicornica* occurs in sandy terraces. Wadis
on chalky ground are dominated by *Salsola tetrandra*.

Slopes between 600 and 1,000 m are bare except for their
wadis which contain the SALSOLETUM TETRANDRAE. Dead
S. tetrandra shrubs which once grew here in a diffused pattern,
can be seen on slopes at 1,000 to 1,200 m. Large areas
between 1,200 and 1,600 m contain many living *Salsola
tetrandra*. The pattern of living and dead shrubs is probably a
result of changes in water regime. At 1,200 to 1,300 m the only
shrubs to survive drought years are found near the few hard-
rock outcrops which store substantial moisture. At lower
elevations characterized by poorer water regimes, even those
shrubs in rock outcrops died. At higher elevations, by contrast,
all microhabitats are supplied with sufficient moisture for the
survival of *S. tetrandra*. The vegetation here at 1,200 to
1,300 m is similar to the vegetation at 700 to 800 m at Gebel
'Ataqa, west of Suez (21). At an elevation of 1,500—1,600 m

in Gebel el 'Igma, north-facing slopes were covered with *Artemisia herba-alba* plants, all of which had died as a result of drought. In a subsequent wet year the area was colonized by *Atriplex leucoclada,* which also invaded the slopes at 1,000 to 1,200 m covered with dead *S. tetrandra. Halogeton alopecuroides,* a xerohalophyte, dominated the south-facing slopes at 1,500 to 1,600 m. *Atriplex leucoclada* could not compete successfully with *Halogeton,* and thus was absent from these sites.

The highest belt of vegetation in Gebel el 'Igma includes the *Chenolea arabica — Atriplex glauca* association on hard chalk. Soft chalk supports SALSOLETUM RIGIDAE on slopes and *Eurotia ceratoides* in wadis. The latter two species are important components of the vegetation of central Asia.

The escarpments of Gebel el 'Igma are covered with very sparse vegetation of the *Salsola tetrandra — Hammada negevensis* association (Fig. 51).

In wet years, areas above 1,300 m are covered with many annuals and geophytes such as *Tulipa polychroma, Leontice leontopetalum,* and *Anthemis melampodina.* The only trees present are several tamarisks. A few wells supply water for Bedouin herds, but there is no agriculture. During rainy periods, the few shallow open cisterns dug by Bedouin, are filled with water. Runoff water from the adjacent reg areas flows to these cisterns through dug channels.

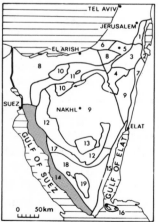

DISTRICT 14.
COASTAL PLAIN OF THE GULF OF SUEZ

Mean annual rainfall—10-30mm;
Mean annual temperature —
22°-24°C;
Number of species — 331;
Area — 7,100 km²
Vegetation mostly contracted;
Vegetation diffused in sandy areas and salt marshes.

DIFFUSED VEGETATION

Association	Substrate
STIPAGROSTIDETUM SCOPARIAE	sand dunes
Astragalus camelorum — Anabasis articulata	sand covering limestone hills
RETAMETUM RAETAMI	deep sand covering limestone hills
ANABASETUM ARTICULATAE ARENARIUM	shallow sand covering alluvial fans
HAMMADETUM SALICORNICAE	sand fields

CONTRACTED VEGETATION

The vegetation in small wadis is dominated by:

Hammada salicornica	*Zygophyllum coccineum*
Anabasis articulata	*Panicum turgidum*
Fagonia mollis	*Schouwia thebaica*
Ephedra alata	*Artemisia judaica*

The arboreal components of large wadis are: *Acacia raddiana, Tamarix aphylla* and *T. nilotica.*

The hydrohalophytes dominating salt marshes along the shore are: *Zygophyllum album, Nitraria retusa, Arthrocnemum macrostachyum, Halocnemum strobilaceum, Tamarix nilotica,* and *Tamarix passerinoides.*

The northern part of the district is strongly influenced by sand intrusions from District 8 and there is no sharp demarcation between the two districts. However, there are differences in vegetation. For example, several Mediterranean species and their associates (e.g., *Scrophularia hypericifolia, Artemisia monosperma, Thymelaea hirsuta*) grow in District 8 but are absent here. On the other hand, *Hammada salicornica* is the dominant semishrub of sandy areas in S. Sinai, but does not occur in District 8.

The sands, which occur in dunes running from north to south, cover limestone hills and alluvial terraces at depths that vary from storm to storm but depend generally on topography. Near Wadi el Haj (15 km WNW of the Mitla Pass) limestone hills without sand cover support only sparse vegetation. Where the sand cover is between 10 and 70 cm, the *Astragalus camelorum — Anabasis articulata* association develops accompanied by other perennials, but no annuals. *Astragalus camelorum* is endemic to this part of Sinai. Semishrubs are killed by sand deeper than one meter. However, *Retama raetam* shrubs survive this sand cover. The ANABASETUM ARTICULATAE ARENARIUM develops on the sand-covered alluvial terraces of Wadi el Haj. Shallow sand on the steep slopes of Wadi el Haj supports *Haloxylon persicum.* Rocks or alluvium completely covered by deep sand are dominated by *Stipagrostis scoparia.*

Most of the area of District 14 is made up of sandy alluvial fans derived from the mountains of Districts 12 and 18. The most common semishrub in the wadis of these fans is *Hammada salicornica,* a psammophyte of extremely hot deserts. In large wadis (e.g., Wadi Sudr and Wadi Gharandal) phytogenic mounds of sand build up around the stems of *Tamarix aphylla* and *T. nilotica* trees; the branches grow above their sand cover and in course of time mounds are formed.

The El Q'a valley, between the magmatic massif of District 18 and the Gulf of Suez, is an extremely arid area. Along the road from Et Tur to Sharm el Sheikh, the distance from one *Hammada salicornica* shrub to the next is measured in kilometers rather than meters. Rain is very infrequent and any flood water is absorbed by the alluvial fans before reaching the valley. In contrast, sections of Wadi Jib'a, Wadi Hebran, and Wadi Isla contain water throughout the year, and support hydrophytes such as *Typha australis, Phragmites australis,* and *Arundo donax.* Strong floods remove parts of these plants and deposit them on stones and shrubs growing in the alluvial fan. *Acacia* trees grow near the wadi mouth, using ground water which penetrates the wadi beds from further upstream. Direct evaporation from the soil surface of superficial underground water has led to the creation of salt marshes populated with halophytes. These salt-resistant plants form belts along the shores and around salt marshes. *Halocnemum strobilaceum* and *Arthrocnemum macrostachyum* dominate the wetter habitats, whereas *Nitraria retusa* and *Zygophyllum album* occur on the drier soils. Many sites with relatively fresh water in this area support wild and cultivated date palms.

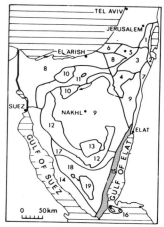

DISTRICT 15.
COASTAL PLAIN AND FOOTHILLS OF THE GULF OF ELAT

Mean annual rainfall — 5-30 mm;
Mean annual temperature — 23°-26°C;
Number of species — 236;
Area — 3,100 km²
Contracted vegetation.

As a result of the great diversity in rock type there are many different plant associations in the wadis of this district. Each of the following species is a dominant in a different association in small wadis: *Hammada salicornica, Abutilon fruticosum, Hibiscus micranthus, Lavandula coronopifolia, Panicum turgidum, Cymbopogon parkerii, Lasiurus hirsutus, Blepharis ciliaris, Eremopogon foveolatus, Solenostemma oleifolium, Zilla spinosa, Gymnocarpos decander, Aerva persica, Artemisia judaica, Launaea spinosa, Seidlitzia rosmarinus, Heliotropium arbainense, Salsola cyclophylla, Capparis cartilaginea, Pterogaillonia calycoptera, Otostegia schimperi, Crotalaria aegyptiaca, Cleome chrysantha, Lindenbergia sinaica, Taverniera aegyptiaca, Cyperus jeminicus.*

The trees found in large wadis are: *Acacia raddiana, A. tortilis, Calotropis procera, Moringa peregrina, Capparis decidua, Leptadenia pyrotechnica, Tamarix nilotica, T. aphylla, Hyphaene thebaica, Salvadora persica.*

The dominant species in the associations surrounding coastal salt marshes are: *Zygophyllum album, Suaeda vermiculata, Nitraria retusa* and *Limonium axillare.* Although *Avicennia marina* is found in coastal salt marshes, it mainly grows in the coastal muddy deposits where it forms the mangroves.

The development of so many different plant associations is possible in this hilly but dry district because of the great variety of rock and soil types.

The most interesting phenomenon of the district is the presence of mangroves on the southern shores of the district. A mangrove was defined by Walsh (128) as a woodland formation below the hightide mark. Mangroves are characteristic of muddy soils in tropical tidewater. The most northerly occurrence of *Avicennia marina* is at the Gulf of Elat, at 28°10' N. This species has been recorded in Japan as far north as 27° N and in South Australia as far south as 37° S (128). The mangroves of Sinai withstand a somewhat cooler climate than those in other areas. According to Walsh (128), mangroves require a mean temperature of not less than 20° for the coolest month of the year and a difference of not more than 5°C between the means of the coolest and the warmest months. In the southern part of the Gulf of Elat the mean temperature for the coolest month is 18.2°C while the temperature difference between the seasons is 14.2°C. Mangroves do not occur in the Gulf of Suez probably because the mean temperature of the coldest month here is only 15°C and the difference between the seasons is 15°C as well.

Mangroves require fine-grained alluvium and shores free of strong wave activity (131), since strong currents would carry away seedlings and destroy the substrate. The mangroves in Sinai occur mainly where a coral reef protects them from strong currents (Fig. 52). The absence of mangroves north of Nabq is probably a result of the low temperatures and the steep cliffs which prevent the development of suitable habitats. *Avicennia marina* trees utilize sea water from the mud in which the roots are anchored. The upper parts of the root grow out of the water and function as respiratory organs (pneumatophores). These pneumatophores absorb air and transport it to the roots under water (Fig. 53). The leaves are coated with a layer of salts excreted by special glands. The seeds are dispersed by the sea water. The tree may reach a height of five meters in the sea but only 0.5—1 m in the coastal

Fig. 52. A schematic substrate-vegetation diagram north of Nabq.

Fig. 53. *Avicennia marina* growing in a dead coral reef showing pneumatophors at low tide.

salt marshes. *A. marina* growing in salt marshes has no pneumatophores.

The vegetation along the coast is arranged in zones nearly parallel to the coast. Figure 52 represents a diagram showing the plant-soil relationships in an alluvial fan near Nabq. *Salvadora persica,* a dominant species in one of the zones, is a prostrate plant that forms sand mounds. (In the Dead Sea Valley this species develops into a tree). Because of prevailing northerly winds, the accumulation of sand takes place on the southern side of the mound, whereas the northern side is constantly eroded (see 35 and Fig. 54). The water table in this zone is several meters deep and the water is fresh.

The zone of *Nitraria retusa — Zygophyllum album* occurs on more saline soil where the water table is closer to the surface than in the *Salvadora* zone. Further down towards the coast is the zone in which the dominant species is *Limonium axillare,* an important halophyte of the Red Sea coasts. Patches of sterile saline soil can also be found in this zone. Occasional specimens of *Avicennia marina* are also found in this zone. Due to the high salinity of the water here, the salt layer excreted by the *A. marina* leaves is thicker than that of the leaves on the trees growing as mangroves in the sea.

The diagram in Fig. 55 represents plant-soil relationships in an area near the shore several kilometers south of that shown in Fig. 52. This area (Fig. 55), which has less muddy shores, a deeper water table and no salt marshes, does not have any mangroves. The wadis in the magmatic or metamorphic hills at the western part of the area support only scant semishrub vegetation. *Lindenbergia sinaica* which is the dominant semishrub in many wadis is surprisingly green even in summer. It is wet to the touch due to a sour-tasting liquid contained in

glandular hairs. This liquid may protect the plant from herbivorous animals. In fact the Bedouin relate that their camels suffer damage to their teeth from eating this plant during extremely dry years when other sources of food are lacking.

Acacia raddiana trees occur in the alluvial fans at the foot of the hills (Fig. 55). They make use of the runoff water coming through the wadis from the magmatic rocks. The density of trees is highest near those mountains contributing the most runoff. Further down the alluvial fans, soil moisture is lower, and there are little chances that *A. raddiana* seedlings will become established. In open areas among the mountains, sometimes characterized as "the moon plains" by visitors, the development of any vegetation depends largely on the rainfall for that particular year. The substrate is fine gravel derived from the disintegration of granite. Precipitation is very infrequent, and green plants of *Pulicaria desertorum* emerge only after storms that sufficiently wet the alluvial fan. This highly scented composite, which can be an annual or perennial, lives only as long as it has water. In other alluvial fans, plants of *Schouwia thebaica,* a crucifer growing up to one meter,

Fig. 54. Plants of *Salvadora persica* showing exposed stems on the northern side of their sand mounds.

Plate 9. The doum palm (*Hyphaene thebaica*) at the oasis of Nuweiba south of Elat. The northernmost specimens of this species are located a few kilometers north of Elat.

develop after a strong rainstorm. It survives longer than the *P. desertorum*. When visiting the arid sites several years after a flood, it is difficult to imagine that conditions were once suitable for the growth of *Schouwia*.

The fossil coral reefs support a more permanent vegetation than is represented by *P. desertorum* or *S. thebaica*. This habitat receives sufficient runoff to support long-living semishrubs and trees. The fine-grained material found in the fossil coral reefs, contributes more runoff than gravels and the wadis here receive more runoff water than wadis in granitic alluvium. In addition, the sand which has accumulated in the wadi channel increases the infiltration of water into the soil. The permanent vegetation in the wadi of the coral reefs causes a local decrease in wind velocity, producing an additional accumulation of sand in the channel. Occasional shrubs of *Leptadenia pyrotechnica* and *Capparis decidua*, growing here in the wadis of the coral reefs, are dominant trees in the wadi vegetation further south along the Red Sea coast (72).

-W-

Teucrium leucocladum & Lindbergia sinaica in wadis

Acacia raddiana in fans near the mountains

Pulicaria desertorum in wadis in rainy years

Cyperus jeminicus — Taverniera aegyptiaca in sandy wadis

-E-

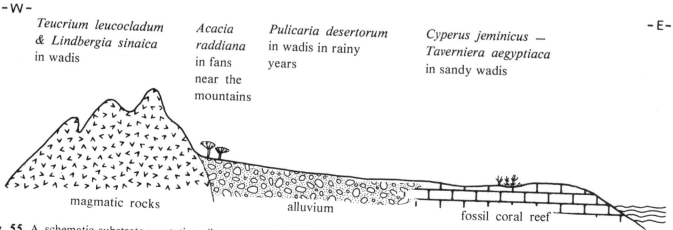

magmatic rocks   alluvium   fossil coral reef

Fig. 55. A schematic substrate-vegetation diagram south of Nabq.

Plate 10. *Cocculus pendulus*, a prominent vine in the warm parts of the Negev and Sinai. Here it is hanging on the cliffs of Wadi Watir, Eastern Sinai.

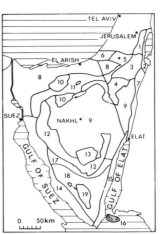

DISTRICT 16.
TIRAN AND SINAFIR ISLANDS

Mean annual rainfall — 5-20 mm;
Mean annual temperature —
25°-26°C;
Number of species — 63;
Area — 130 km²
Vegetation in a contracted pattern.

The dominants in the wadi associations are:

Salsola cyclophylla	*Anabasis setifera*
Taverniera aegyptiaca	*Cyperus jeminicus*
Seidlitzia rosmarinus	*Salvadora persica*, and
Suaeda vermiculata	*Capparis cartilaginea*
Zygophyllum coccineum	

The dominants in the belts of salt marshes are: *Suaeda vermiculata, S. monoica,* and *Limonium axillare.*

These very small arid islands support few plants. The three major substrates in Tiran are: gypsum rock in the main mountain range, small alluvial fans in the coastal plain, and fossil coral reefs (Fig. 56). The vegetation on the gypsum outcrops is very sparse, consisting of occasional *Capparis cartilaginea* shrubs and *Salsola cyclophylla* semishrubs. The wadis draining the mountain support *Salvadora persica* trees and the tropical vine *Pentatropis spiralis.* Sandy wadis on the coral reefs are similar to those south of Nabq (District 15) and support a similar vegetation. The *Salsola cyclophylla — Cyperus jeminicus* association sparsely covers most of the alluvial fans. Denser vegetation develops in the coastal salt marshes. These marshes are drier than those in District 15 but similarly support *Limonium axillare.* Wadis draining these fossil coral reefs support the famous rose of Jericho (*Anastatica hierochuntica*).

Sinafir is the smaller of the two islands and has even sparser vegetation. *Suaeda vermiculata* grows sporadically near the shores and one central salt marsh is dominated by *Arthrocnemum macrostachyum* and *Suaeda monoica.*

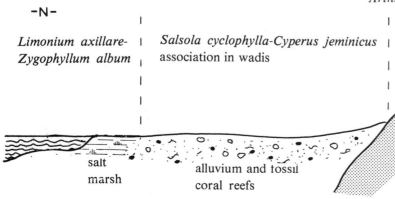
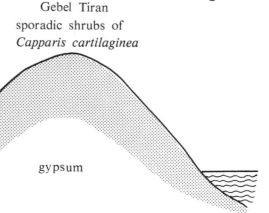

-N-

Limonium axillare-
Zygophyllum album

Salsola cyclophylla-Cyperus jeminicus
association in wadis

Gebel Tiran
sporadic shrubs of
Capparis cartilaginea

-S-

salt
marsh

alluvium and fossil
coral reefs

gypsum

Fig. 56. A schematic substrate-vegetation diagram in Tiran Island.

Plate 11. Aerial view of the transition between the Sandstone Belt and the Lower Sinai Massif. Note the presence of *Acacia raddiana* trees at the margins of large wadis and in the center of small wadis in the area of magmatic rocks and their absence from the sandstone zone.

DISTRICT 17.
SANDSTONE BELT

Mean annual rainfall —
30-50 mm;
Mean annual temperature —
17°-22°C;
Number of species — 238;
Area — 2,100 km²
Vegetation pattern varies
according to edaphic conditions.

DIFFUSED VEGETATION

Association	Substrate
Mosaic of IPHIONETUM MUCRONATAE and	bedded sandstone
SALSOLETUM CYCLOPHYLLAE	massive sandstone
Launaea spinosa — Pituranthos triradiatus	massive sandstone
GYMNOCARPETUM DECANDRI	basalt, granite and other
VARTHEMIETUM MONTANAE	magmatic rocks
Hammada salicornica — Ephedra alata	loose sand covering other strata
Hammada salicornica — Heliotropium digynum	loose sand covering other strata
Retama raetam — Heliotropium digynum	loose sand covering other strata
Ephedra alata — Convolvulus lanatus	loose sand covering other strata
Ephedra alata — Anabasis articulata	loose sand covering other strata
HALOXYLETUM PERSICI	loose sand covering other strata

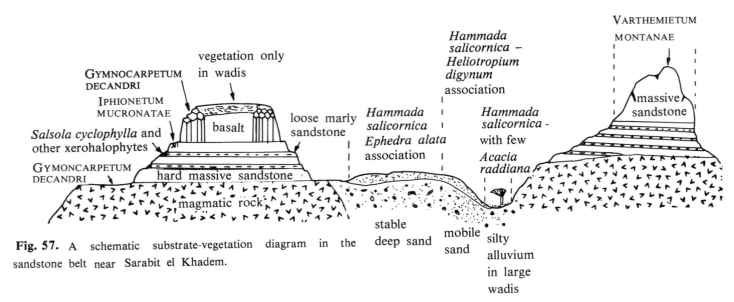

Fig. 57. A schematic substrate-vegetation diagram in the sandstone belt near Sarabit el Khadem.

Association	Substrate
Gymnocarpos decander — *Convolvulus lanatus*	loose sand covering other strata

CONTRACTED VEGETATION

The dominants in wadis with sandy ground are:

Hammada salicornica	*Retama raetam*
Ephedra alata	*Iphiona scabra*
Anabasis articulata	*Zilla spinosa*

The dominants in wadis with silty ground are:

Salsola cyclophylla	*Reaumuria hirtella*
Artemisia herba-alba	*Anabasis setifera*
Gymnocarpos decander	

The vegetation here depends on the particular type of rock and its way of weathering. Hard sandstone interbedded with softer layers containing clay or silt results in a step-like topography. At the foot of each step there is a small slope of debris derived from the softer rock. IPHIONETUM MUCRONATAE, a lithophytic community, dominates rock crevices and fissures of the hard rock. Plants growing in the softer rock have less available water and more salt, and are mostly xerohalophytic semishrubs at 800 m to 1,200 m elevation. Since rainfall decreases with decreasing elevation, the vegetation of the soft rock layers disappears at a higher elevation than that of the hard rocks. Columnar outcrops of the basalt layer (Fig. 57) interbedded in the sandstone support GYMNOCARPETUM DECANDRI. The steep slopes form steps that are the optimal habitat for *Gymnocarpos decander* (cf. Fig. 25). Exposed magmatic rocks of district 18 (that are overlain in district 17 by sandstone) support the same plant community. Soil derived from basalt or other magmatic rocks has a high silt and clay content and can support vegetation only in wadis (Plate 1).

The silty alluvial plains among the sandstone hills do not support vegetation unless overlain by sand. Several communities dominated by *Hammada salicornica*, *Ephedra alata*, and other species cover the sand in a diffused pattern. Mobile sands that mainly accumulate on steep slopes of large wadis support the *Hammada salicornica* — *Heliotropium digynum* association. New stems from exposed roots and new roots can develop from sand-covered stems of this *Heliotropium*. *Acacia raddiana* trees occur in magmatic sections of large wadis that extend from the sandstone belt to the lower Sinai massif. However, those trees are nearly absent in the sandstone sections of the wadis (Plate 11), since the sandy alluvium does not contain sufficient moisture. It is evident from Plate 11 that in large wadis *A. raddiana* is restricted to the margins of the main wadi channel. In smaller tributaries the trees are more widely dispersed because the erosive force of flood water does not uncover the roots. Silty alluvial fans near the cliffs of Gebel et Tih support associations of *Artemisia herba-alba* at high elevations and of *Anabasis articulata* at lower elevations.

Massive sandstone with few or no interbedded soft strata has a water regime similar to that of smooth-faced limestone or magmatic rocks. Massive sandstone in Sinai supports VARTHEMIETUM MONTANAE at 900 to 1,200 m elevation. Several rare species found in this plant community include: *Pennisetum elatum*, *Polygala sinaica*, *Ferula sinaica*, *Sageretia brandrethiana*, and *Rhamnus dispermus*. The *Lauanea spinosa* — *Pituranthos triradiatus* association occurs at somewhat lower elevations. At Timna, near sea level (District 7), the same sandstone supports no vegetation.

Plate 12. Date palms (*Phoenix dactylifera*) growing spontaneously in a wadi 20 km northeast of Et Tor, southern Sinai.Together with *Juncus arabicus*, these date palms indicate that superficial underground water can be found here.

DISTRICT 18.
LOWER SINAI MASSIF

Mean annual rainfall —
30-50 mm;
Mean annual temperature —
15°-22°C;
Number of species — 407;
Area — 4,650 km²
Vegetation both diffused and
contracted

CONTRACTED VEGETATION
Common dominants in small wadis are:

Hammada salicornica	*Retama raetam*
Aerva persica	*Zilla spinosa*

Artemisia judaica	*Tephrosia apollinea*
Fagonia mollis	*Ochradenus baccatus*
Hyoscyamus boveanus	*Gomphocarpus sinaicus*
Otostegia schimperi	*Iphiona scabra*
Capparis cartilaginea	*Moricandia sinaica*
Launaea spinosa	

Trees in large wadis are: *Acacia raddiana, Calotropis procera* and *Tamarix nilotica*.

Trees near springs are: *Moringa peregrina, Ficus pseudosycomorus,* and *Phoenix dactylifera*.

DIFFUSED VEGETATION

Association	Substrate
Iphiona mucronata — Echinops glaberrimus	smooth-faced rocks
Varthemia montana — Globularia arabica	smooth-faced rocks
Artemisia herba-alba — Echinops glaberrimus	fissured hard rocks
Gymnocarpos decander — Farsetia aegyptiaca	fissured hard rocks
Anabasis articulata — Fagonia mollis	plains of granitic gravel

Fig. 58. Date palms in Wadi Feiran oasis.

This district is a transition area between the hot and coastal plains (District 14 and 15) and the relatively cool area of District 19. In District 18 we find cold-demanding plants and moisture-demanding plants growing at their lowest altitude in Sinai. The thermophilous species *Acacia raddiana* and *Hammada salicornica* do not grow above 1,300 m. *Achillea fragrantissima*, an important component of the wide wadi vegetation in District 19, is present here mostly near District 19. At the lowest elevations, vegetation is restricted to wadis. Then, proceeding upwards, the general pattern consists of: semishrubs near large rocks and in crevices of smooth-faced outcrops; diffused vegetation on fissured rocks and the diffused *Anabasis articulata — Fagonia mollis* association in granitic gravel plains. This last association is relatively drought-resistant and is replaced by *Artemisia herba-alba* communities at higher elevations.

Especially significant here in District 18 are the several oases in large wadis that drain District 19. Wadi Feiran oasis (Fig. 58) is famous for its date palms. The water table is high here, and the dense Bedouin population subsists on dates, various fruit trees, irrigated vegetables, and the large herds of black goats that graze on the diffused vegetation of the higher mountains. Throughout the district there are many smaller oases similarily farmed by Bedouin.

Many wadis leading to the Gulf of Suez have springs and for most of the year flowing water. These include Wadi Isla, Wadi Hebran, and Wadi Jib'a all three wadis support hydrophytes wherever there is sufficient soil.

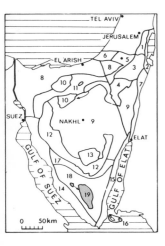

DISTRICT 19.
UPPER SINAI MASSIF

Mean annual rainfall — 70-100 mm;

Mean annual temperature — 9°-15°C;

Number of species — 419;

Area — 1,300 km²

Vegetation mostly diffused.

Association	Substrate
Gymnocarpos decander — Artemisia herba-alba	Fissured rocks and stony soil
Artemisia herba-alba — Zilla spinosa	Fissured rocks and stony soil
Artemisia herba-alba — Fagonia mollis	Fissured rocks and stony soil
Artemisia herba-alba — Anabasis setifera	Fissured rocks and stony soil
Artemisia herba-alba — Halogeton alopecuroides	Fissured rocks and stony soil

Plate 13. *Hyoscyamus boveanus*, a poisonous semishrub of southern Sinai which contains high quantities of atropine, hyoscyamine and hyoscine.

Association	Substrate
Artemisia herba-alba — Atraphaxis spinosa	Fissured rocks and stony soil
Artemisia herba-alba — Stachys aegyptiaca	Fissured rocks and stony soil
Artemisia herba-alba — Tanacetum santolinoides	Fissured rocks and stony soil
Gymnocarpos decander — Otostegia schimperi	Smooth-faced granite
Globularia arabica — Verbascum decaisneum	Smooth-faced granite
Tanacetum santolinoides — Phlomis aurea	Smooth-faced granite
Tanacetum santolinoides — Astragalus echinus	Smooth-faced granite
Varthemia montana — Pistacia khinjuk	Smooth-faced granite
Varthemia montana — Iphiona mucronata	Smooth-faced granite
Achillea fragrantissima — Artemisia judaica	Wide valleys

Dominant plants near the springs are: *Holoschoenus vulgaris, Schoenus nigricans, Mentha longifolia, Tamarix nilotica* and *Phoenix dactylifera*.

In small dripping springs on cliffs the dominants are: *Hypericum sinaicum* and *Adiantum capillus-veneris*.

This district is the coolest in the region owing to its high elevation. The flora is diverse and includes Irano-Turanian, Mediterranean and Sudanian plants that are isolated from their main areas of distribution. The closest station for several of the isolated species here are in Iran or on Mt. Hermon (Anti-Lebanon). *Crataegus sinaica* and *Scrophularia libanotica* occur both in Sinai and on Mt. Hermon.

Primula boveana (Plate 14) is a rare endemic which has been isolated in Sinai since the Tertiary (130). Its nearest relatives are found in eastern Africa, Yemen, and the Zagros

Fig. 59. Gebel Serbal (2,070 m) with *Acacia raddiana* trees in a wadi at 900 m in the foreground.

Plate 14. *Primula boveana,* and endemic species of the Upper Massif of Southern Sinai developing in small springs at high elevations.

Mountains of Iran. There are several other endemic species most of which are restricted to smooth-faced rock outcrops

The flora of this district is dominated by Irano-Turanian species, and the most common plant is *Artemisia herba-alba* It is accompanied by *Gymnocarpos decander* in fissured rocks at lower elevations, and by *Zilla spinosa* and *Fagonia mollis* in stony alluvium. *Anabasis setifera, Halogeton alopecuroides,* and *Atraphaxis spinosa* are associates in so derived from dark volcanic rocks. *Stachys aegyptiaca* an *Tanacetum santolinoides* accompany *Artemisia* at the foot c smooth-faced rock outcrops at low elevations and on ston slopes at higher sites.

Rock vegetation here is very rich in semishrubs and tree and poor in annuals. Characteristic trees and shrubs includ *Crataegus sinaica, Pistacia khinjuk, Ficus pseudosycomoru: Cotoneaster orbicularis, Sageretia brandrethiana, Rhu tripartita,* and *Rhamnus dispermus.*

The two most prominent rock types in the highlands c southern Sinai are old black volcanic rocks with man fissures and red or white granite with few fissures. Rain wate falling on the volcanic rock is distributed among man fissures. The rain water falling on the red granite concentrate in a few fissures which provide a favorable water regime a these sites. The fissured volcanic rocks of Gebel Katherin (2,640 m) do not support a single *Pistacia khinjuk* tre

Fig. 60. Smooth-faced granite at the peak of Gebel Beida.

whereas thousands of these trees occur on Gebel Serbal (2,070) m) which consists of smooth-faced red granite (Fig. 59). Gebel Beida at 1,750 m also supports *P. khinjuk* trees (Fig. 60). This illustrates that rock structure has a greater influence on the distribution of vegetation than does elevation.

Flowing water, waterfalls and pools are common all over the area with red granite. The hard rocks in these mountains contain soft dykes in which large quantities of water accumulate giving rise to many small springs. The monks of the monastery, as well as the Bedouin, have dug many wells here in order to irrigate their many fruit trees and vegetable beds. The gardens include old trees of pear, carob, apple, walnut, almond, peach, fig, and olive, as well as grape vines. The Bedouin of the Jabaliya tribe are keen farmers and always try to improve by learning from any possible source. They graft pears onto *Crataegus sinaica* trees. They built their terraces in sites having a high water table and irrigate their vegetables and trees from a shallow well in the garden.

Annual plants develop every year as weeds in the gardens, growing on the fine-grained irrigated soil. Other sites support annuals only in rainy years and mainly on soils rich in fine-grained particles. Outcrops of hard rocks are poor in annuals. The distribution of annuals reflects the localized nature of the rainstorms. The most common annuals are *Paracaryum intermedium, Lappula sinaica, Gypsophila viscosa, Eremopoa persica, Boissiera squarrosa* and *Ziziphora tenuior.*

As discussed above, each district has typical flora and vegetation. The environmental conditions and the history of the flora have a substantial impact on the number of species growing in each area. To compare the floristic composition of the whole desert area discussed here with other areas and to summarize this chapter, species diversity analysis is presented below.

PLANT SPECIES DIVERSITY IN THE DESERTS OF ISRAEL AND SINAI
PARAMETERS OF SPECIES DIVERSITY

According to Preston (106), the number of plant species in an area is related to the degree of species diversity of the area and to its size, as follows:

$$S = KA^z$$

or in its logarithmic form,

$$\log S = \log K + z \log A$$

where:

S = number of species
K = a constant dependent on the degree of species diversity of the region
A = area of the region
z = rate of species increase with area.

The number of species in our deserts are compared with other regions of the world by plotting curves of species versus area on a log/log scale (Fig. 61). The calculated number of species/log area curves (z) for most regions are given in Table 1.

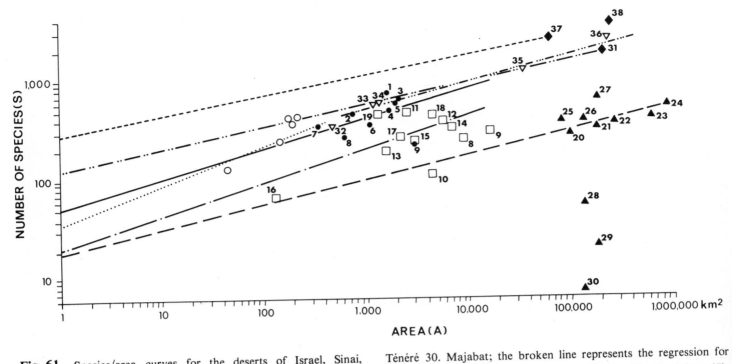

Fig. 61. Species/area curves for the deserts of Israel, Sinai, Sahara and other areas in the world. Dots with figures 1-9 (●———) denote the districts of Israel; squares with figures 8-19 (□—·—) denote the districts in Sinai; (▲———) = areas of the Sahara (after 107): 20. south Moroccan Hammada 21. south Tunisian Sahara 22. Spanish Sahara 23. Mauritania 24. central Sahara 25. Tassilis 26. Ahaggar 27. Tibesti 28. Djourab 29.

Ténéré 30. Majabat; the broken line represents the regression for areas 20-24. (○··········) some California Islands (after 67); (—··—·) British Isles 31. Britain (♦) total flora (after 68); (▽···—·) Israel (see Table 2), 32. Mt. Carmel 33. Judean Hills 34. Upper Galilee 35. Judean Hills and southern Sharon 36. Israel (total); (-------) mainland areas of California (after 67) 37. California coast (total) 38. Arizona (total, after 74).

IMPACTS OF ENVIRONMENTAL STRESS ON K

The species diversity (K) of a region, is influenced by the environmental stress which in deserts affects the number of micro-habitats available for plants. The main environmental factor limiting growth in the desert is obviously water. The following examples show the effect of varying amounts of rainfall on otherwise similar habitats. Sand fields 15 km south of Be'er Sheva with a mean annual rainfall of 150 mm, support 20—30 species per 20 square meters, while sand fields near Tasa in northern Sinai with a mean annual rainfall of 60—70 mm, support only 5—7 species per 20 square meters. No plants grow in similar sandfields of central Sahara (4) with less than 20 mm of annual rainfall. In Lisan Marl of the Dead Sea Valley many annual species occur where mean annual rainfall is over 100 mm, but none where rainfall is under 50 mm. The increase in number of species and number of habitats supporting plants as the result of increasing moisture at higher elevations was discussed in detail under Districts 18 and 19.

THE IMPACT OF PLANT GEOGRAPHY ON K

The species diversity of a given area is also influenced by the proximity of different plant geographical regions. In transition zones, species of more than one phytogeographical region may grow together. Mobile dunes of the northern coast of Israel are dominated by *Ammophila litoralis* while dunes of northern Sinai are dominated by *Stipagrostis*

scoparia. In the area of Ashqelon — Gaza, both species grow in the same habitat.

The chorotype of a plant (see pp. 35-36) represents its area of distribution. The chorotypic diversity of an area is low when most of the species there are of the same chorotype. If the species of an area are evenly distributed among several chorotypes, the chorotypic diversity is high. A numerical expression of chorotypic diversity is given by the Shannon-Wiener equation: $H = -\sum Pi \log_e Pi$

where $Pi = \dfrac{\text{number of species of the chorotype concerned}}{\text{total number of species}}$

The exponential expression e^H represents the chorotypic diversity of each district. The lowest values of e^H are expected in the center of the phytogeographical regions and the highest values in transitional areas. The results given in Table 2 indicate that the Negev and Judean deserts have greater chorotypic diversity than the Sinai. Districts with a low e^H value (i.e. below 8) cover 95 percent of Sinai and only 14 percent of the deserts of Israel. The high diversity in districts 19 and 11 in Sinai is a result of the many outcrops of smooth-faced rocks that provide refuge for species which penetrated the area in the past. According to studies in Egypt (11), in the Sahara (108), in the Sudan (132), in Israel (66, 93) and northern Sinai (52), the climate of the Middle East and North Africa during the Late Pleistocene was

Table 1. Number of species, areas, z and K values
for various regions of the world.
See text for explanation of symbols.

Locality	Size of area (km²)	No. of species (S)	z	calculated No. of species per km² (K)	– source of data
Israeli deserts	13,200	1,130	0.299	50	
Sinai	61,100	812	0.307	21	28
Sahara			0.2214	20	107
California Islands			0.37	39	67
British Isles	229,850	1,666	0.209	125	68
Netherlands	34,800	1,357	0.28	73	88
Israel	30,500	2,307*	0.264	145	147
Arizona	294,873	3,370			74
California coast	63,479	2,325	0.158	276	67

* Information obtained from this source relates only to S.

Table 2. Chorotypic diversity (e^H) of the deserts of Israel and Sinai.
The calculations of e^H are based on data from Fig. 16.

Negev and Judean Deserts District	e^H	Sinai District	e^H
1	7.76	8 (Sinai portion)	9.85
2	11.69	9 (Sinai portion)	7.07
3	9.44	10	4.75
4	8.45	11	8.86
5	8.15	12	8.33
6	9.23	13	6.13
7	9.74	14	8.07
8 (Negev portion)	7.39	15	7.64
9 (Negev portion)	8.08	16	4.12
Total	9.94	17	7.25
		18	8.66
		19	9.83
		Total	10.40

considerably moister than at present. Species adapted to the moist climate have survived in appropriate microhabitats in rock crevices.

The low number of species per square kilometer in the Sahara (K = 20 species) and in Sinai (K = 21 species), reflects their harsh environments when compared to the milder deserts of Israel (K = 50). The meeting of four phytogeographical regions in the Israeli deserts may also contribute to the higher K values.

THE VALUES OF z

A high value of z indicates a steep increase in number of species with increasing area. The main factor influencing the distribution of plants in deserts is the water regime which is determined by climate, topography and substrate. The more types of rock and kinds of topography in a given area, the greater the diversity in water regime. If an area contains one large geomorphological structure (such as an anticline, syncline, etc.), the diversity of habitats will be much lower than in an area of the same size containing many small geomorphological structures. The geomorphological structures of the Sahara are huge, those of Sinai are large, while those of Israel are relatively small. Also isohyets are the densest in the desert areas of Israel and the least dense in the Sahara (91). As a reflection of both geomorphological and rainfall patterns, the z value of the Israeli deserts is 0.299, that of Sinai is 0.307, and in the Sahara it is only 0.221.

To conclude, the total number of species in the deserts of Israel and Sinai is influenced by the steepness of climatic gradients, the size of geomorphological structures, the meeting of several phytogeographical regions, and the refuge provided by rocks for previously prevailing flora. The combination of all these result in a high number of species in Sinai, lower in the Israeli deserts, and the lowest in the Sahara.

Acacia raddiana in **Sinai**.

CHAPTER 4. CHARACTERIZATION OF SOME PROMINENT OR RARE SPECIES

This chapter describes 47 of the nearly 1,300 species recorded from the Negev and Sinai. Some are the dominants of many plant communities, while others have interesting patterns of distribution. Plants are grouped according to the most significant environment in which they occur in the deserts of Israel and Sinai. The groupings are not always mutually exclusive.

The moisture regime of each habitat is characterized as follows: a) Optimal — the species grow in a diffused pattern and may even dominate the vegetation on various types of soil and slope; b) Dry — drier areas in which the species is restricted to wadis or soil pockets in rocks i.e., where moisture conditions are more favorable than in the surrounding areas; and c) Wet — areas in which the species is restricted to drier habitats than most of the surroundings. The map accompanying the description of each species indicates its distribution pattern following the above classification. In some cases complete information is lacking but we were able to delimit the areas where the plant grows in a diffused pattern. In other cases we could only delimit the area where the species occurs without specifying its pattern of distribution

LEGEND OF THE DISTRIBUTION MAPS:

area with optimal water regime (diffused pattern on various soil types and slopes)

area where the plant grows in a diffused pattern

relatively dry area where the plant is restricted to habitats wetter than most of the area (contracted pattern)

area where the plant is known to occur but distribution pattern is not specified

relatively wet area where the plant is restricted to habitats drier than most of the area

▼   one or a few individuals

The chorotype (see p. 36) designation is taken from Zohary (144, 145), Grünberg (54), Feinbrun (44), and Täckholm (121). The Bedouin plant names listed here are based on the study of Bailey and Danin (5). Information concerning the meaning of many of the scientific names is taken from Smith (120). Plant names are listed as follows: Latin scientific name, family name, common English name (where applicable), Hebrew name in transliteration, and Bedouin name (where applicable) in transliteration.

The following hints may give an idea of how the Hebrew and Bedouin names are pronounced:

a as a in the Spanish word padre or as o in pop as pronounced by Americans.

u as u in put.

o as o in bone but without any diphthong.

e as e in let.

ee as ee in feet.

kh as j in the Spanish word junta or ch in the German name Bach.

dh as th in this.

' represents a guttural sound typical to Semitic languages. It may be approximated by a "catch" in the throat.

TYPICAL DOMINANTS IN AREAS WITH DIFFUSED VEGETATION ON HARD ROCKS

Artemisia herba-alba Asso
COMPOSITAE

English: wormwood, white wormwood.
Hebrew: la'anat hamidbar.
Bedouin: shih.
Name: The genus is named after Artemis, the goddess of modesty and of the hunt in ancient Greece. This was probably the common name of a particular species in Greece or in Rome and was adopted by Linnaeus as a scientific name for the whole genus. The specific epithet *"herba-alba"* which means "the white herb" describes the white woolly stems and leaves.

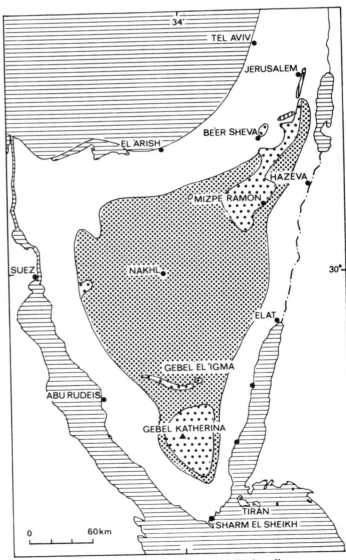

Fig. 62. Distribution map of *Artemisia herba-alba*.

Chorotype: West Irano-Turanian.

Distribution (Fig. 62): *Artemisia herba-alba* is found in Israel and Sinai at elevations from 250 m below sea level to 2,640 m above sea level. In grows as a dominant in a diffused pattern in areas of the Judean Desert, Northern and Central Negev highlands, Gebel et Tih, and southern Sinai Massif at elevations above 1,500 m. In these areas it occupies slopes of weathered limestone, dolomite, granite, magmatic and metamorphic rocks, as well as sandy-silty soils. In Israel and Sinai the area in which it grows in wadis (contracted pattern) is even larger than the area where it grows in a diffused pattern. It dominates in the wadis draining the gravel plains of central Sinai between Elat-Nakhl and Mitla pass. In the rest of the relatively dry area (Fig. 62), wormwood is found as a component of drought-resistant associations such as those in which *Zygophyllum dumosum* or *Anabasis articulata* is dominant. In the very dry region of Makhtesh Hazera, where *Z. dumosum* dominates the slopes, wormwood grows in rocky wadis or in crevices of smooth-faced dolomite. Some individuals are even found near the Dead Sea growing in large canyons draining the Judean Desert. In areas more humid than in the optimal area, wormwood dominates chalk outcrops near Jerusalem and north of Be'er Sheva. Plants of the sub-Mediterranean batha which accompany *A. herba-alba* here are less drought-resistant than the companions of this plant in its optimal area.

Growth: At the beginning of the winter, buds situated at the base of branches which carried flowers last year elongate into stems (102). The growth rate in winter is slow and it increases towards spring. The upper parts of the stems start branching during April and May. Inflorescences with minute tubular florets develop on these branches from August to November. In December one can still find flowering individuals. Achenes are dispersed by wind from November to January. The leaves which grow on the stems in winter and spring are large and are dissected into many lobes. In the axils of the large leaves and at the upper parts of the stems, there develop smaller leaves with fewer lobes densely covered with white hairs. Wormwood starts shedding its winter leaves in April. The leaves at the base of the plant are shed first (101). After fruit ripening, the upper parts of the stems which carried the inflorescences become dry. In the next season, if the plant has sufficient water, it will develop new stems. In drought years, the large winter leaves fail to develop and the summer foliage remains on the plant without any renewal of stems and branches. Neither will it bloom in such a year. After several years with insufficient rain, some sections of the plant may die without influencing the stems and roots of the other sections (Fig. 63 and p. 30). This mechanism is effective in the optimal area where the plant inhabits rocky or stony habitats. Here the heterogeneous soil may supply enough water to at least some sections of the plant nearly every year. In the wadis of the gravel plains, west of Elat, the plants of *A. herba-alba* grow on sandy-silty ground which absorbs and holds high quantities of water during floods. This water is sufficient for the plant to live one to two years. If there are several drought years in succession, a common

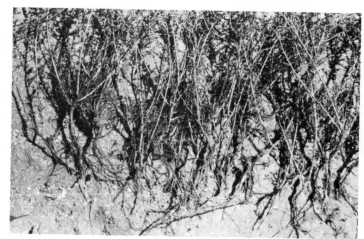

Fig. 63. *Artemisia herba-alba* shrub that was artificially split into subunits.

phenomenon in the gravel plains, most wormwood plants in the wadis die. Since these wadis contain no rocks, there are no fissures in which water can concentrate. In these wadis, wormwood becomes re-established after drought years from seeds rather than from surviving sections of older plants.

Uses: The green leaves and stems, boiled in water with sugar, produce a very tasty drink. It can also be used to flavor tea. The drink is used in folk medicine for the relief of upset stomach. Inhaling the vapor of boiling water containing leaves and stems of the wormwood may ease colds. The white woolly insect galls developing on the stems may be used to start a fire without matches. This is done by concentrating the sun's rays with a magnifying glass or by using flint with steel to produce a spark on the gall (see p. 127).

Anabasis articulata (Forssk.) Moq.
CHENOPODIACEAE

English: articulated anabasis, jointed anabasis.
Hebrew: yafruk hamidbar.
Bedouin: ajram.

Name: The scientific name of the genus is derived from Greek meaning "without basis". It is not clear why it was so named. The specific epithet and the Hebrew name refer to the articulated stems of the species.

Chorotype: Saharo-Arabian extending into the Irano-Turanian region.

Distribution: (Fig. 64): *A. articulata* grows as a dominant in a diffused pattern on several different soils — on limestone outcrops on the 'Avdat plateau, Hatira Mountains, Har Sagi, Gebel Yiallaq, and the Gebel et Tih plateau; hard chalk in the Judean Desert; Neogene sand on the plains of Rotem and Yeroham — Dimona; stable sand and sand covering loess or gravel plains between Revivim and Suez; loess soil near Sede Boqer. In all of the above areas, the mean annual rainfall is 80—150 mm and the elevation is 300—1000 m. In areas with low precipitation it is confined to high elevations and in areas with relatively high precipitation it dominates at low elevations.

In southern Sinai, *A. articulata* grows as a dominant at elevations of 1200—1400 m in various soils derived from

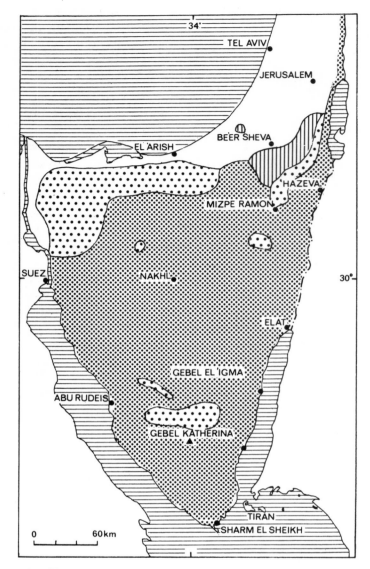

Fig. 64. Distribution map of *Anabasis articulata*.

area where the plant grows in a diffused pattern

relatively dry area where the plant is restricted to habitats wetter than most of the area (contracted pattern)

relatively wet area where the plant is restricted to habitats drier than most of the area

magmatic rocks. The areas where *A. articulata* dominates in diffused vegetation are more arid than those dominated by *Artemisia herba-alba*. In areas wetter than optimum, *Anabasis articulata* grows sparsely and is not dominant. In both the Negev and in Sinai it is common as a component in associations dominated by *Artemisia herba-alba*. In those large areas of the southern Negev and central Sinai where the vegetation is contracted, *A. articulata* dominates in wadis of chert gravel plains. It dominates there wherever the wadi slope is very gentle and the ground is filled with loess and gravel. It also dominates in the wadis of conglomerates and other rock types.

Growth: The succulent stems start their growth in spring, branch in summer, and bear flowers in September and October.

The diaspore develops from a flower and contains one seed surrounded by the ovary wall and five sepals, each of which develops a wing-like structure after fertilization. These diaspores, are discharged during December and January, are dispersed by wind. The immature wings vary in color from cream or white to purple and differ from one shrub to the other. The diversity in wing color is higher among populations in the optimal area than in drier areas. After diaspore dispersal, the stems they were growing on desiccate and fall. The seeds germinate easily during the first rains. When conditions for germination are not adequate, the seeds lose their germinability in a few weeks. In dry years the plant may not bloom at all, and then the stems desiccate because of drought and not because of the blooming cycle.

Zygophyllum dumosum Boiss.
ZYGOPHYLLACEAE
English: bushy bean caper, shrubby zygophyllum.
Hebrew: zugan hasiakh
Bedouin: 'adhbee.
N a m e : The leaf is built up of a cylindrical petiole with a pair of leaflets. This is suggested by the generic name. The specific epithet assigned by Edmond Boissier refers to the shrubby and lignified nature of the species when compared to other species.
C h o r o t y p e : East Saharo-Arabian.
D i s t r i b u t i o n (Fig. 65): The bushy bean caper grows in a diffused pattern on rocky and stony slopes of limestone and dolomite at elevations ranging from 300 m below sea level near the Dead Sea to nearly 1,000 m above sea level at Gebel et Tih in Sinai. At Har Zavoa, the northern boundary of the plant in the northern Negev, it grows on a sandy layer where the water regime is relatively poor compared with that of other rock types in the area. On the Rahama Mountains it occupies relatively dry habitats such as southern slopes, loessial colluvium and slopes of stony soil with dark chert stones. In the central Negev (District 4) it grows on the tops of flat hills and on southern slopes. All the above sites are wetter than the optimal for *Z. dumosum*. However, it succeeds in these sites only in relatively dry microhabitats. In areas where the mean annual rainfall is 70 to 100 mm and the elevation is below 500 m, it grows on limestone hills on various types of slopes. Where the mean annual rainfall is less than 50 mm, it is restricted to relatively wet habitats such as old wadi beds and conglomerate plateaux. On limestone plateaux it is restricted to those slopes which receive additional runoff water. Dry areas with relatively soft limestone support the bean caper only in wadis, especially where a hard layer is exposed in the channel of the wadi. Many areas in Districts 11 and 12, which are the limestone mountains of Sinai, support *Z. dumosum* in a diffused pattern in the hard rocks and in a contracted pattern in the soft strata. Both the hard and soft rocks may occur in the same mountain.

Zygophyllum dumosum does not grow on magmatic or metamorphic rocks. This may be due to the absence of several

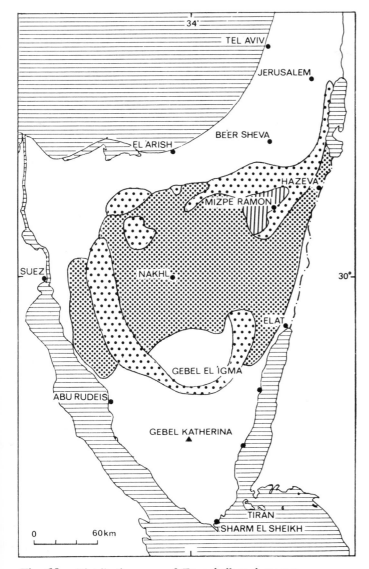

Fig. 65. Distribution map of *Zygophyllum dumosum*.

area where the plant grows in a diffused pattern

relatively dry area where the plant is restricted to habitats wetter than most of the area (contracted pattern)

relatively wet area where the plant is restricted to habitats drier than most of the area

minerals which are present in the limestone and dolomite rocks.

G r o w t h : The bean caper branches are covered in summer with leaf petioles one to three years old (Fig. 66). These petioles are small and wrinkled but swell to 2 to 3 times their size after the first effective rain. A few days after this first rain, new stems develop from dormant buds. If the rainfall is limited, the bud will produce a stem containing one node with a pair of leaves and one flower. More rainfall will induce formation of more such stem units, each with a pair of leaves and a flower. In each instance the new stem develops from the axil of the most recently formed leaf pair. If there is shortage of water, there will be no new growth and the plant will remain only with its old petioles for another year. By summer,

all leaflet pairs become dry and are often shed. If there is a sufficient supply of water, the shrubs will remain green in summer and even bloom.

The epidermis of the leaf petiole is built up of one layer of cells in winter and becomes multicellular in summer (38). If the shrubs are artificially supplied with water in summer, then the epidermis remains single-layered. When there is a deficiency of water, the plant decreases its transpiration rate by decreasing the transpiring area and by developing a multilayered epidermis on its petioles. The shedding of leaflets may lead to a 96% decrease in total leaf weight (102). Apart from the decreased leaf area, the whole plant may become drastically reduced in size (see pp. 30 — 31 and fig. 20). In subsequent good years the shrub may return to its former size. This adaptation of the bean caper to seasonal and annual changes in rainfall is reflected in the existence of 200 to 300 year-old shrubs.

The fruits of *Z. dumosum* are five-winged containing one or more seeds in each wing. The fruit wall contains salt which functions as germination inhibitor (76). The salt delays germination until leached by a strong rain which ensures favorable conditions for the establishment of the seedlings.

Fig. 66. *Zygophyllum dumosum* in its typical habitat. Branch in winter (left) with live pairs of leaflets and in summer (right) with shriveled leaflets and sausage-shaped live petioles.

Thymelaea hirsuta (L.) Endl.
THYMELAEACEAE

English: shaggy sparrow-wort, hairy thymelaea.
Hebrew: mitnan sa'ir.
Bedouin: mitnan.

Name: The generic name in Latin is derived from Greek and refers to the resemblance of the leaves to thyme leaves. The Hebrew name is taken from the Arabic. Since the fibers of this plant are made into ropes, a suggestion was made that the plant be named "yitran" in Hebrew, "yeter" meaning a rope. The suggestion was based on the biblical story of Samson and Delilah where the word "yetarim" is used for the ropes which bound Samson (22).

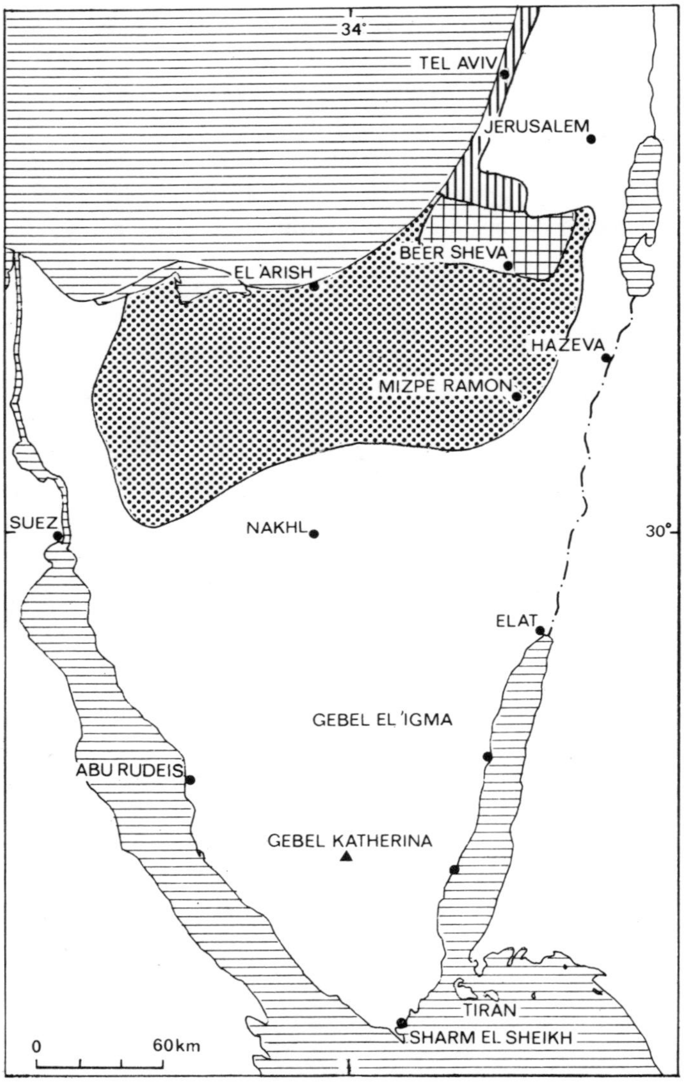

Fig. 67. Distribution map of *Thymelaea hirsuta*.

relatively dry area where the plant is restricted to habitats wetter than most of the area (contracted pattern)

area with optimal water regime (diffused pattern on various soil types and slopes)

relatively wet area where the plant is restricted to habitats drier than most of the area

Chorotype: Mediterranean and Saharo-Arabian.

Distribution (Fig.67): *Thymelaea hirsuta* grows in a diffused pattern on various soils in the northern part of Districts 5 and 6 where the mean annual rainfall is 180 to 300 mm. This seems to be the optimal area of the species in this part of the world. In drier areas, east and south of the optimal area, it is found in relatively wet habitats such as in wadis and at the foot of smooth-faced limestone outcrops. It is also common on shallow sand covering silty soil; the sand allows water to infiltrate efficiently and the silt holds the water well. In the same area, on silty soils it grows only in wadis. In areas receiving 70 to 150 mm of rainfall, it dominates in loessial wadis where the flood water is distributed over much of the width of the wadi. However, in wadis with gravelly ground it grows only sparsely. In wetter areas north of the Negev it grows in soils in which the water regime is comparatively poor. For example, it grows on sandy soils along the Mediterranean coast, in chalky-marl outcrops on Mt. Carmel, and in similar habitats in Mediterranean zones of Europe (e.g. Greece and Spain).

Growth: The seedlings have relatively large bluish leaves which are glabrous and spreading. The lower side (abaxial) of the small green adult leaves are shiny and glabrous and have no stomata. The upper leaf side (adaxial), which is hairy and contains stomata, is closely appressed to the stem.

Massive germination of *Thymelaea hirsuta* seeds takes place once every several years. This plant becomes established quickly since it is resistant to grazing. Even the black goat, which is known to eat nearly everything, avoids *T. hirsuta*. It is a shrub (phanerophyte) which can be up to 2 m high (Fig. 68). It may bloom during much of the year. It is dioecious; i.e., the male and female reproductive organs develop on different individuals.

Uses: Strong ropes can be prepared from the bark fibers (Figs. 69, 70, 149, 150). Avoid putting the fibers into the mouth as they contain harmful substances.

Fig. 68. *Thymelaea hirsuta*.

Fig. 69,70 Rope of *Thymelaea hirsuta* fibers made into a carrier for water-jugs.

Noaea mucronata (Forssk.) Asch. & Schweinf.
CHENOPODIACEAE
English: thorny saltwort, pointed noëa.
Hebrew: noeet kotsaneet.
Bedouin: sirr.
Name: The genus is named after the 19th century French botanist Noé who studied the Labiatae of northwest Africa and the Canary Islands. The name "sirr" is also applied by the Bedouin of southern Sinai to another spiny semishrub *Atraphaxis spinosa*.
Chorotype: West Irano-Turanian.
Distribution (Fig. 71): In the vicinity of Be'er Sheva *N. mucronata* grows as a dominant in a diffused pattern on the slopes with several soil types. In most areas where *Artemisia herba-alba* dominates in a diffused pattern, *N. mucronata* accompanies it. The distribution patterns of these species indicate that *A. herba-alba* is more drought-resistant. In Gebel et Tih, *A. herba-alba* grows from elevations of 500 m upwards, but is accompanied by *N. mucronata* only from elevations of 1,000 m upwards. In Gebel el 'Igma, *A. herba-alba* prevails in the range of 500 to 1,600 m, whereas *N. mucronata* is found only in the range

of 1450 to 1600 m. In the semi-steppe batha bordering the Mediterranean and desert habitats where *Sarcopoterium spinosum* is dominant, there are many individuals of *Noaea*, but *A. herba-alba* is absent. This batha may be considered to be wetter than the optimal for *N. mucronata*. Far from the desert boundary, in areas of the Mediterranean maquis 27 km southwest of Jerusalem, *N. mucronata* is found as a dominant on chalk outcrops. Near Hebron it grows in overgrazed areas which once were covered by maquis and are now nearly devoid of any competing plant species. In areas drier than the optimum *Noaea* is found in soil pockets of rock outcrops; it only rarely occurs in wadis. It does not grow on magmatic or metamorphic rocks.
Growth: The life cycle of *N. mucronata* is similar to that of many semishrubs in the desert (Fig. 12). In the beginning of winter, small twigs (temporary brachyblasts) bearing summer

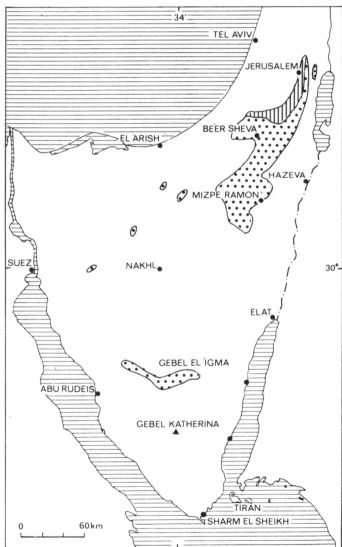

Fig. 71.   Distribution of *Noaea mucronata*.

area where the plant grows in a diffused pattern

relatively wet area where the plant is restricted to habitats drier than most of the area

Fig. 72. A two-year old stem (dolychoblast) of *Noaea mucronata* in winter. F = dry spiny twigs (brachyblasts) which carried last year's flowers and fruits; S = small summer leaves near base and W = winter leaves of a new stem (dolychoblasts).

leaves produce new stems (Fig. 72). The section developing in winter has longer leaves and internodes than the section that had developed in summer. In the course of the season the same new stems develop smaller and smaller leaves. In summer minute leaves develop in the axil of the long winter leaves. The winter leaves are shed in summer. The upper branches of the new stems become spiny, bear flowers and then fruits. Each fruit develops five wings from the back of the sepal. Each diaspore is composed of one seed enclosed in a thin-walled fruit, subtended by five winged sepals.

Sarcopoterium spinosum (L.) Spach
ROSACEAE
English: prickly shrubby burnet, thorny burnet.
Hebrew: seerah kotsaneet.
Bedouin: natsh.
Chorotype: East Mediterranean.
Distribution (Fig. 73): *S. spinosum* is common throughou the Mediterranean phytogeographical territory of Israel. I dominates the batha vegetation in rocky areas at the margin and in deforested areas in this territory. *S. spinosum* appear to be an important component of the batha climax vegetatio near the margins of the Mediterranean territory. However, disappears from areas where the maquis or forest returr

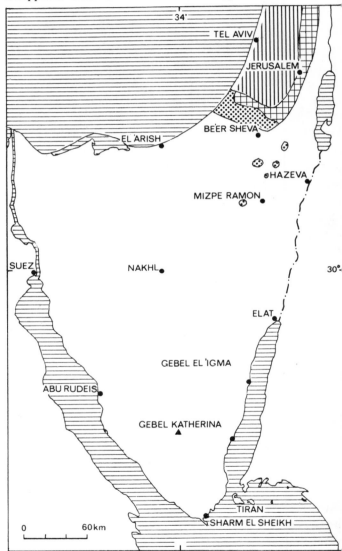

Fig. 73. Distribution map of *Sarcopoterium spinosum*.

area with optimal water regime (diffused pattern on various soil types and slopes)

relatively dry area where the plant is restricted to hat tats wetter than most of the area (contracted pattern)

relatively wet area where the plant is restricted to habitats drier than most of the area because *S. spinosum* is sensitive to shade. Presumably, its primary habitat was the semi-steppe batha; and after the destruction of the primary forests and maquis, it penetrated considerable areas of the Judean Mountains, Carmel, Samaria and Galilee (86). *S. spinosum* is a dominant mainly in rocky areas, whereas in deep clayey soils (grumusols) it is rare. In several of the ridges of the Negev, *S. spinosum* grows in crevices of smooth-faced limestone and dolomite and in wadis draining slopes containing such rocks.

The dispersal unit is a fruit with a spongy outer part covered by an epidermis almost impermeable to water. Seeds are dispersed short distances by wind and longer distances by floating on flood water. However, there is no connection between water courses draining the southern Judean Mountains and the Negev Highlands. Presumably, *Sarcopoterium spinosum* has survived in the Negev as a relict of a period a few thousand years ago, when a wetter climate prevailed in the Negev. Perhaps, *S. spinosum* grew at the periphery of the maquis of *Quercus calliprinos,* pollen grains of which were discovered in prehistoric sites (66).

Growth: The life cycle of *S. spinosum* differs slightly from that of the typical chamaephyte (102). In spring new twigs (dolychoblasts) develop from buds situated at the base of the stems. These twigs do not develop flowers during their first year and terminate in several spines that become dry towards summer. Small summer leaves develop in the axils of the winter leaves. The following winter new twigs develop at the base of last year's stems. The buds in the upper part of these stems develop into short reproductive branches (brachyblasts). These have a rosette of leaves and a stalk bearing flowers and fruits. The reproductive branches function for two years. The flowers are unisexual; the inflorescences consist of all male flowers, or all female flowers or a mixture of both types. The plant is monoecious and is pollinated by wind. The fruit, 4—5 mm in diameter develops from an epigynous ovary and resembles an apple in shape. However, instead of being juicy, the outer part has a spongy tissue which functions in dispersal.

Phlomis brachyodon (Boiss.) Zoh.
LABIATAE
Hebrew: shalhaveet keetsrat-sheenayim.
Name: The scientific generic name is derived from the common Greek name. The Hebrew name refers to the yellow color of the corolla in many species of this genus. The specific epithet refers to the short teeth of the calyx.
Chorotype: West Irano-Turanian.
Distribution (Fig. 74): This species is endemic to the semi-steppe batha on both sides of the Jordan Valley. A few individuals are found in soil pockets and fissures in rocks of the Judean Desert and in the northern Negev. These individuals in the deserts are probably relicts of a period

when a wetter climate prevailed in the area. The plant is a dominant in chalky soils with approximately 250 mm of annual rainfall. It accompanies *Sarcopoterium spinosum* in the semi-steppe bathas. In areas with 200 mm of rainfall, the plant is restricted to relatively wet habitats, such as northern slopes, wadis, and rock outcrops.

Growth: The small summer leaves are thickly covered with stellate hairs (Fig.16). The hair layer on each side of the leaf is 1—2 times as thick as the leaf proper. New growth starts in winter when longer stems develop from the short ones with summer leaves. The winter leaves are larger and their hair cover is thinner (Fig 16). In spring, yellow flowers develop at the upper parts of the long stems. While the plant is in bloom, the winter leaves begin to dry and summer leaves develop. The diaspore is a fruiting calyx containing four nutlets and the dry corolla acts as a wing for wind dispersal.

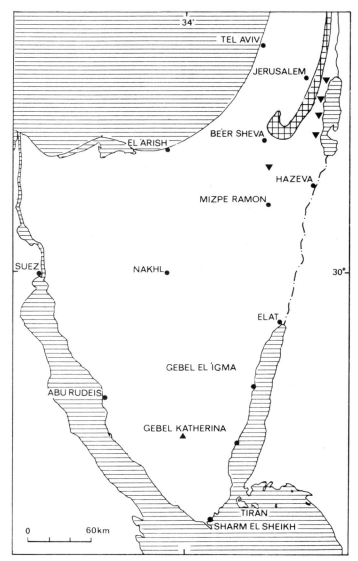

Fig. 74. Distribution map of *Phlomis brachyodon*.

area with optimal water regime (diffused pattern on various soil types and slopes)

▼ one or a few individuals

Pituranthos tortuosus (Desf.) Benth.
UMBELLIFERAE
Hebrew: kazuakh 'akum.
Bedouin: zagukh, gazukh.
Name: The scientific generic name was applied by Viviani in 1842, who described a species with large bracts and small umbels which look like flowers. In Greek *pituron* means cover or bracts, and *anthos* means a flower. The Hebrew name is derived from the Bedouin name.
Chorotype: East Saharo-Arabian.
Distribution (Fig. 75): *P. tortuosus* grows as a dominant in a diffused pattern in small areas near the semi-steppe bathas of Districts 5 and 6. Occasional individuals are found in rocky habitats and wadis in those sections of the Negev Highlands which are drier than the optimum (Fig. 75). In

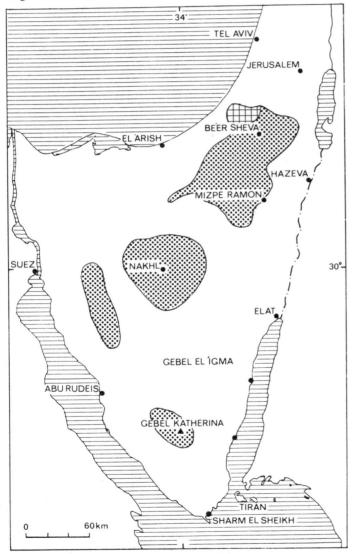

Fig. 75. Distribution map of *Pituranthos tortuosus*.

area with optimal water regime (diffused pattern on various soil types and slopes)

relatively dry area where the plant is restricted to habitats wetter than most of the area (contracted pattern)

central Sinai it is a dominant in a few wadis in the gravel plains. In the granites of southern Sinai it is found in fissures of smooth-faces outcrops at high elevations. However, at lower elevations *Pituranthos triradiatus* grows in the granite outcrops. *P. triradiatus* does not reach dominance in our area and is found as an occasional species companion in the wadi vegetation of the Negev and Sinai deserts. *P. tortuosus* has a strong fragrance when the green stems are crushed, whereas *P. triradiatus* does not have a smell.
Growth: In winter new stems sprout from the base of the green stems from the previous year. Young stems have leaves with a broad sheath covering the whole stem. The later leaves near the top of the stem are smaller and have fewer lobes than the basal leaves which develop early in the winter.
The leaves which develop in summer have no lobes but only sheaths. In summer all sheaths become dry. During most of the year the green stems function in photosynthesis as described for stem assimilants (p. 29). Stems which start growing in winter become branched towards summer and bear umbels at their tips. Flowers of *P. tortuosus* can be found all through the year but most of the blooming takes place at the beginning and end of the summer.
Uses: The young leaves of *P. tortuosus* are palatable, their taste resembling that of celery. A soup can be prepared from the green stems and leaves of *P. tortuosus*. Some Bedouin use this plant to prepare a hot drink.

PLANTS OF LOESS SOILS

Anabasis syriace Iljin
CHENOPODIACEAE
English: Syrian anabasis.
Hebrew: yafruk tlat-kanfee.
Bedouin: 'adhu.
Chorotype: West Irano-Turanian.
Distribution (Fig. 76): In the valleys and badlands of the northern Negev *A. syriaca* dominates on loess soils in areas with 150—250 mm of rainfall at elevations of 200 to 600 m. In the Central Negev Highlands it is found near the isohyet of 100 mm at elevations of 700 to 1000 m. It also grows in habitats rich in nitrogen salts such as goat and cattle corrals, possibly due to its tolerance of saline conditions (27). In a given area *A. syriaca* grows equally as well on loess soil superficially plowed by Bedouin as on unplowed loess.
A. syriaca is not found on loess soil where the salts are sufficiently leached by rain water. This is probably a result of competition from glycophytes (23, 27). At the arid margins of its range *A. syriaca* is restricted to terraces located just above the present water channel in the wadi. The plant is not common in wadis, and is absent from sandstone and from the magmatic and metamorphic rocks of southern Sinai.
Growth: The distal parts of stems are dry in winter. The bark is green in summer and yellowish-brown in winter. The bases of stems contain renewal buds which begin to grow in spring. These stems branch in the summer during flowering

(Fig. 77). The fruits are dispersed in October and November when the bark desiccates, turning most of the stems brown. In this species the bark is green and photosynthetically active mainly during the dry summer and not in the wet winter. In drought years flowering may not occur and there is a reduction in growth and in the transpiring area. Summer growth and activity is possible because the plant is able to use salty water in the deeper soil layers. This water is not available to the winter annual species. Very few other perennials accompany *A. syriaca* at sites where it is dominant.

Fig. 77. *Anabasis syriaca* in summer. P = previous year's stem; S = this year's stems; F = flower bearing twigs.

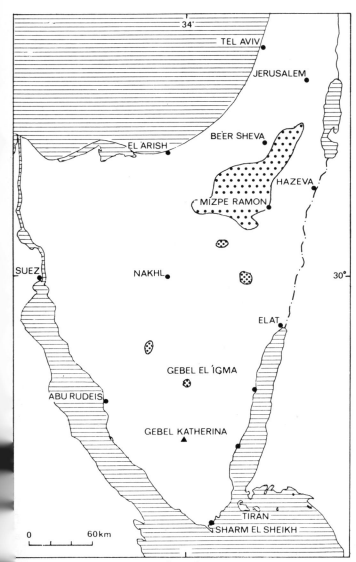

Fig. 76. Distribution map of *Anabasis syriaca.*

area where the plant grows in a diffused pattern

relatively dry area where the plant is restricted to habitats wetter than most of the area (contracted pattern)

Hammada scoparia (Pomel) Iljin
CHENOPODIACEAE
Hebrew: khammadat hamidbar.
Bedouin: hidad.
Name: Several different Latin names have been applied to this plant by different authors. One such name was *Haloxylon articulatum* (Cav.) Bunge. From Zohary (144) we learn that Bunge included in *Haloxylon* what Cavanilles (1794) named *Salsola articulata*. Since Forsskal (1775) applied the name *Salsola articulata* to what we now call *Anabasis articulata*, Cavanilles was incorrect in giving the name *Salsola articulata* to *Hammada scoparia*. Pomel (1875) who studied the North African flora named our taxon *Haloxylon scoparium*. Iljin (1948), who divided the genus *Haloxylon* into two genera based on floral morphology, gave the plant its present name. Not all authors have accepted Iljin's nomenclature and this species is also named *Haloxylon scoparium* and *Arthrophytum scoparium*.
Chorotype: West Irano-Turanian and Saharo-Arabian.

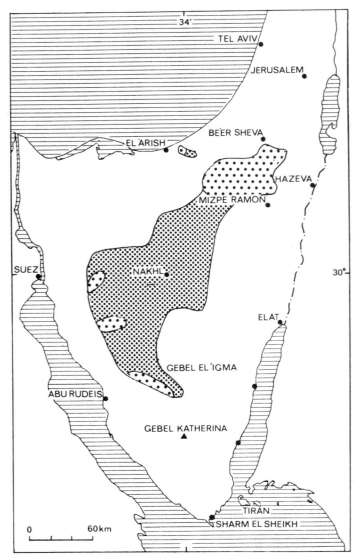

Fig. 78. Distribution map of *Hammada scoparia.*

   area where the plant grows in a diffused pattern

relatively dry area where the plant is restricted to habitats wetter than most of the area (contracted pattern)

Distribution (Fig. 78) : *H. scoparia* grows in a diffused
pattern in the Negev Highlands, Gebel er Raha, Gebel
Budhiya, and Gebel et Tih. It is rather common in the wadis
of central Sinai. In the northern Negev (District 3) it is found
in various different habitats including loess soils, stony slopes
of chert rocks, limestone slopes, and marl outcrops. Between
Yeroham and Nizzana it is restricted to loess soils, where it is
accompanied only by a very few other semishrubs. *H. scoparia* like *Anabasis syriaca* can also make use of salty
water in summer. These two species are among the very few
semishrubs which can tolerate the water regime of salty loess
soils.

H. scoparia also grows in nitrogen-rich sites, such as the ancient Nabatean towns of 'Avdat and Nizzana. In non-saline
soils (i.e., leached loess in areas with more than 200 mm of

rainfall, or sandy soils in which water infiltrates readily) these
two species lose their competitive advantage and are rarely
found. *H. scoparia* is even more drought-resistant than
A. syriaca and hence is found in drier areas.

In District 9 (central Sinai), *H. scoparia* is common in the
wadis of gravel plains. The silty soil in these wadis also
supports *Artemisia herba-alba. H. scoparia* also dominates
the loess flood plains of large wadis such as Wadi El 'Arish,
Wadi Quraya and Wadi el Bruk. The water regime of these
flood plains resembles that of small wadis. *H. scoparia* is not
common in wadis in the Negev.

Growth: New branches begin to develop in spring. During
July and August, lateral stems with short internodes develop
at the distal part of these spring branches. The stems with
short internodes produce flowers in late summer and ripe
fruits in autumn. The diaspores, consisting of seeds, fruit wall,
and winged sepals, are dispersed by wind in winter. Prior to
ripening, the color of the sepal wings varies considerably. The
lateral stems which carried the flowers and fruits desiccate
during the winter, the period of low activity for this plant.
During drought years, *H. scoparia* does not bloom and loses
internodes from the existing branches.

The structure of the root system is influenced by the type
of habitat (122). On loess plains roots penetrate to a depth of
20-30 cm; at the margins of gravelly wadis — to a depth of
over 2 meters; and in sandy loess the roots may penetrate as
much as 3 m. In rocky habitats root development probably
varies according to the site characteristics.

Erodium hirtum Willd.
GERANIACEAE

English: hairy heron's-bill, hairy stork's-bill.
Hebrew: makor hakhaseeda hasaeer.
Bedouin: tumayer, bilbis.

Name: The ripe fruit resembles a stork's head (the ovary)
and stork's beak (a column developed in the flower's center).
In Greek *erodios* means stork and *hirtum* means hairy. The
Hebrew name means hairy stork's beak. The Bedouin name
refers to the root tubers of this species.

Chorotype: Saharo-Arabian.

Distribution (Fig. 79). This hemicryptophyte occurs in
most of the Negev, Judean Desert, and Sinai where the mean
annual rainfall ranges from 30 to 300 mm. It inhabits stony
and arid loess soils. It is absent in sand and in salty soils. It is
most abundant in the loess plains and on the slopes of
limestone hills in Districts 3 and 4.

Growth: Plants more than one year old develop leaves
following the first effective rains. In wet years the plants
develop many stems, leaves, and flowers. The dark color at
the base of the petals (Fig. 80) attracts pollinating insects to
the stamens and styles at the center of the flower. The ripe
ovary, composed of five carpels, splits into five diaspores
(Fig. 81), each containing one seed. The ovary wall covering
the seed is hairy and has a sharply pointed tip. The part of

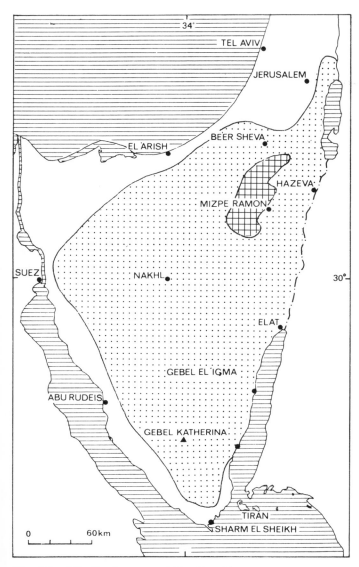

Fig. 79. Distribution map of *Erodium hirtum*.

area with optimal water regime (diffused pattern on various soil types and slopes)

area where the plant is known to occur but distribution pattern is not specified

Fig. 80. Flowers of *Erodium hirtum*. The contrasting colors attract pollinating insects to the reproductive organs at the center of the flower.

the beak closest to the seed twists as a result of changes in humidity. Strong winds dislodge the ripe diaspores and carry them away from the plant. The diaspores penetrate the soil like a screw as a result of the above-mentioned twisting of the beak. Ants collect and consume large numbers of *E. hirtum* seeds. Those left in the soil germinate in winter, each one developing a small tuber on its tap root. The leaves desiccate in summer. In the following years the root system branches and additional tubers develop at the tips of the lateral roots. Tubers are red-brown on the outside. Young tubers are white inside, but with age the inside turns red-brown and the tuber becomes lignified. Porcupines dig out and eat these tubers during the summer.

Uses: During the summer the young tubers are juicy and sweet and can be eaten. When the tuber becomes red inside, it is bitter and no longer palatable.

Fig. 81. Five diaspores of *Erodium hirtum* developing from one flower. Each diaspore has a "feather" which functions in wind dispersal.

Fig. 82. A typical landscape of *Suaeda asphaltica* plant community in the Judean Desert.

PLANTS OF CHALK AND MARL

Suaeda asphaltica. (Boiss.) Boiss.
CHENOPODIACEAE
English: asphaltic sea blight.
Hebrew: ukam midbaree.
Bedouin: suwayda.

Name: The genus was named by Forsskal (1775) who studied the flora of Egypt. He used the Bedouin vernacular name which means "the black". The Hebrew name also means black. The plant becomes black when the leaves dry before falling. The species name was given by Edmond Boissier (1853) who collected the plant near the Dead Sea, then known as Mare Asphaltitis.

Chorotype: Saharo-Arabian. Endemic to Israel and Jordan.

Distribution (Fig. 83): *S. asphaltica* grows mainly on steep slopes of chalk, marl and phosphorite. In Israel it is restricted to areas with a mean annual rainfall of 70 to 100 mm. In the northern Judean Desert it dominates areas between 150 m above sea level and 150 m below sea level. Further south, *S. asphaltica* dominates chalk slopes at higher elevations, e.g. 500 to 600 m at Sede Boqer. At the eastern margin of its range and south of 'En Gedi, it is found on north-facing slopes and in wadis. At the western margins of its range it is restricted to east and south-facing slopes. At the center of its range in the northern Judean Desert, it grows on all the slopes. Therefore, the optimal range of this species appears to be northern Judean Desert.

The plant is adapted to a specific level of salinity, and can not compete with other species east of its belt. Near the surface the soil between the *Suaeda* shrubs is leached. This allows many species of glycophytes to become established and prevent the growth of *Suaeda* seedlings. However, these seedlings do become established in rock outcrops where salinity is high. Mature *Suaeda* shrubs also create a favorable microhabitat for their own seedlings. Excess salt absorbed by the plants is diluted and retained in the succulent leaves. The

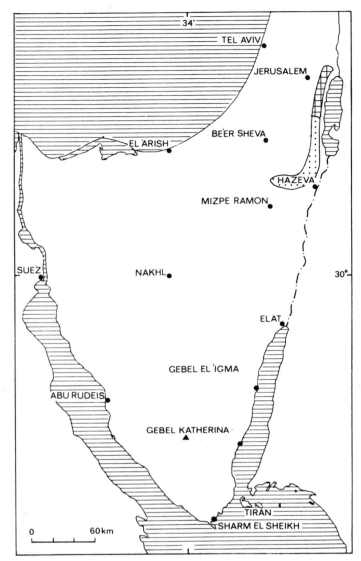

Fig. 83. Distribution map of *Suaeda asphaltica*.

area with optimal water regime (diffused pattern on various soil types and slopes)

area where the plant is known to occur but distribution pattern is not specified

germination until most of the salts are leached out. Soon after seed dispersal, the leaves dry and turn black. The shrubs often appear as black dots on an otherwise white landscape (Fig. 82).

Hammada negevensis Iljin & Zohary
CHENOPODIACEAE

Hebrew: khammadat hanegev.
Bedouin: samur, suwaydi hidad.
Name: The generic name was discussed under *H. scoparia*. The species, described from specimens collected near Sede Boqer, was named after the Negev. The Bedouin name refers to the dark or black color of the plant.
Chorotype: East Saharo-Arabian. Endemic to Israel and Sinai.

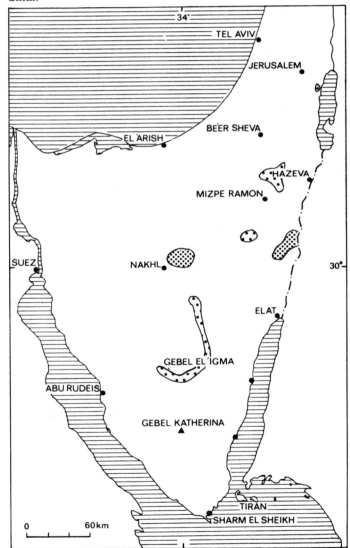

Fig. 84. Distribution map of *Hammada negevensis*.

area where the plant grows in a diffused pattern

relatively dry area where the plant is restricted to habitats wetter than most of the area (contracted pattern)

leaves fall and slowly decay. The salt thereby released creates a highly saline environment near the base of the shrub. When electrical conductivity is used to measure salinity, we find that the surface soil in areas between shrubs has a value of 1mMhos/cm as compared with 8mMhos/cm directly under the *Suaeda* shrubs (27). Transition area have intermediate values. Whereas glycophytes are incapable of growing in the saline soil near mature *Suaeda* shrubs, *Suaeda* seedlings can become established near either a living or a dead shrub.
Growth: The shrubs are leafless in summer. After effective rains, new stems develop at the base of the previous year's stems. The leaves and flowers develop on the new stems. Because the pedicels and petioles are joined, the flowers appear to grow from the petioles. The one-seeded fruits subtended by sepals are dispersed at the beginning of summer. The succulent sepals containing salt may retard

Fig. 85. *Hammada negevensis* growing in wadi east of Sede Boqer.

Distribution (Fig. 84): *H. negevensis* grows as a dominant in a diffused pattern on the steep slopes of the 'Avdat Plateau and Gebel el 'Igma. It occurs in wadis at drier locations such as Zin Valley, 'Arava Valley and near Nakhl in Sinai (Fig. 85).

Growth: This species differs from the two other species of *Hammada* listed in this book in having leaves which are cylindrical and up to 1 cm in length. It blooms in autumn and the winged diaspores are dispersed by wind in winter. Populations of this species in the Negev have light cream-colored sepal wings, whereas in Sinai there are also specimens with pink to violet wings.

Salsola tetrandra Forssk.
CHENOPODIACEAE
Hebrew: milkheet kaskesaneet.
Bedouin: feers.
Name: The Latin generic name and the Hebrew name refer to the plant's affinity for saline soils. Forsskal (1775) used the epithet *tetrandra* because he found four stamens in the flowers of his type specimen. This is in contrast to most species in the genus which possess five stamens.
Chorotype: Saharo-Arabian.
Distribution (Fig. 86): *Salsola tetrandra* grows as a dominant in a diffused pattern in areas of chalk and marl having 50 to 100 mm annual precipitation. It is found at altitudes of 350 m below sea level near the Dead Sea, at elevations of 300 to 600 m above sea level in the eastern part of District 3, and at 1,200 to 1,600 m at Gebel el 'Igma. In drier regions it is a dominant in wadis (Fig. 87). *S. tetrandra* is the most drought-resistant xerohalophyte growing on chalk and marl in Israel and Sinai.
Growth: Even though no growth occurs in summer, the small scale-like leaves are still green and densely hirsute. Stems (dolychoblasts) begin to elongate in winter and continue growing during the spring, at which new lateral stems (brachyblasts) develop in the axils of last year's leaves.

The growth of dolychoblasts continues during the spring while new brachyblasts develop at axils of last year's leaves. Flowers are produced on the brachyblasts along the whole length of the dolychoblasts.

Salsola tetrandra flowers in February at lower elevations near the Dead Sea but not until June at the high elevations of Gebel el 'Igma. Most flowers have four stamens and four sepals. In spring 1969, we studied many plants of *S. tetrandra* near the southwestern shore of the Dead Sea. At this site each plant examined had both tetramerous and pentamerous flowers. Brachyblasts which already existed in the summer of 1968 bore only pentamerous flowers, while brachyblasts produced in the winter of 1969 bore only tetramerous flowers. In 1970 and again in 1974, the same plants bore only tetramerous flowers. The tetramerous and

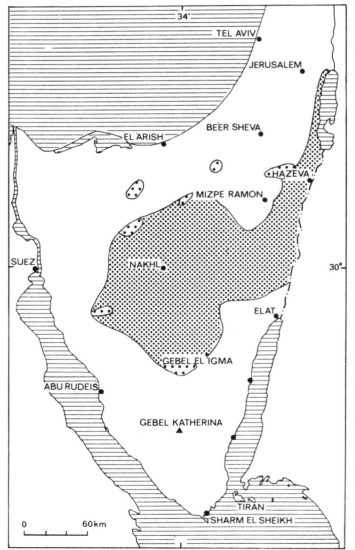

Fig. 86. Distribution map of *Salsola tetrandra*.

 area where the plant grows in a diffused pattern

relatively dry area where the plant is restricted to habitats wetter than most of the area (contracted pattern)

Fig. 87. *Salsola tetrandra* as a dominant in a wide wadi split into many water channels near Nakhl, central Sinai.

pentamerous flowers of *S. tetrandra* both produce viable seeds. Since the same plants had both kinds of flowers in one year and only tetramerous flowers in other years, it is suggested that climatic influences are involved here.

There has been some taxonomic confusion between *S. tetrandra* and its North African relative *S. tetragona*, although both are distinctive species. Many authors have erroneously regarded the unwinged diaspores of *S. tetrandra* as sterile or male forms of *S. tetragona*.

Two months after pollination those parts of the brachyblasts that bore flowers become dry. On the other hand, green leaves on dolychoblasts or brachyblasts which did not bear flowers remain alive during the summer. The fruits are dispersed in summer when the brachyblasts break at the nodes into diaspores. The diaspore consists of an internode with two fruits subtended by sepals and bracts. The sepals and bracts prevent the seed from germinating under unfavorable conditions. The bracts, succulent when green but dry in the diaspore, contain salts which retard germination until leached by strong rains. Mature plants have scale-like opposite leaves. Seedlings and young plants have leaves up to 5 mm long which are nearly opposite. Leaf characteristics are influenced by environmental conditions. Plants growing in saline soil have hairy and distinctly succulent leaves. However, progeny grown in Jerusalem from seeds collected in Gebel el 'Igma were found to have leaves that were much less succulent and hairy than their parents. Many studies show that the succulence of halophytes increases with increased soil salinity (126).

PLANTS OF SANDY HABITATS

Stipagrostis scoparia (Trin. & Rup.) De Winter
GRAMINEAE
English: triple-awned grass.
Hebrew: mal'anan hamatateem.
Bedouin: sabat.
N a m e : The Latin generic name is derived from Greek; *stuppe* means tow (the fiber of flax) and *agrostis* means a kind of grass. Until recently, the genus *Aristida* included all species with three awns. In 1963 De Winter suggested dividing the genus as follows: *Aristida* to include species with unfeathered awns, e.g., *A. coerulescens*, *A. pumila*, and *A. sieberiana*; *Stipagrostis* to include species having awns with lateral hairs, e.g., *S. raddiana*, *S. scoparia*, *S. plumosa*, and *S. obtusa*. The genus *Stipagrostis* was described in 1832, but was generally ignored until De Winter's study.
C h o r o t y p e : Saharo-Arabian.
D i s t r i b u t i o n (Fig. 88): *Stipagrostis scoparia* inhabits mobile sands usually in pure stands. Its optimal range is difficult to delimit on a map because its distribution depends on sand mobility as well as moisture regime. In the mesic coastal Mediterranean area north of Ashqelon, the mobile

sands are inhabited by *Ammophila litoralis* rather than *S. scoparia,* which occurs only sparsely.

Growth: *S. scoparia* germinates abundantly only following several consecutive rainy days. Under these conditions the seedlings become established rapidly (Fig. 89). Those nodes which are covered with sand produce roots and above-ground stems. As the density of *S. scoparia* increases, the sand movement in the area is curtailed, making conditions unfavorable for its further growth. The stems and roots of *S. scoparia* have a relatively short life span, and new sand cover is required for the development of additional roots and stems. Without a continual cover of sand *S. scoparia* plants die (see pp. 19 and 51). Thus, *S. scoparia* does not merely tolerate sand cover, it requires it.

Bedouin in northeastern Sinai graze their goats on *S. scoparia* and cut the stems to produce fuel for bread baking in El 'Arish. As a result, the density of *S. scoparia* is substantially decreased and the sands remain mobile. The

Fig. 89. *Stipagrostis scoparia* growing in mobile sands in Sinai.

reduced vegetation cover near the Israeli border can easily be identified by satellite imagery (Fig. 8). An area of northwestern Sinai closed to Bedouin from 1967 to 1975 has become noticeably revegetated with *S. scoparia* — the dominant plant in the sand dunes.

In mobile sands, *S. scoparia* grows and flowers throughout the year. In semistable dunes vegetation cover is higher and *S. scoparia* flowers only in spring. The awned diaspores, carried by the wind, are deposited on the leeward side of dunes where the wind velocity is much reduced. Wind-blown sand is also deposited at these sites, covering the diaspores. Many seeds are dispersed in the dunes during storms and germinate following sufficient rainfall.

Artemisia monosperma Del.
COMPOSITAE
English: single-seed wormwood.
Hebrew: la'ana khad-zar'eet.
Bedouin: 'adhir.

Name: For a discussion of the generic name see *Artemisia herba-alba* (p. 76). Raffaneau Delile, the botanist with Napoleon's expediton to Egypt, named the species. He noted that each flowering head contains 10 androgynous florets and two female florets, only one of which ripens. Even though the name la'ana appears in the Bible, the identification of *Artemisia* with the Biblical la'ana is not accepted by many scholars of Biblical plants.

Chorotype: Saharo-Arabian and Mediterranean.

Distribution (Fig. 90): *A. monosperma* occurs mainly in the sands of northern Sinai and the Coastal Plain of Israel. It is hard to delimit its optimal range because, apart from annual rainfall, sand texture and mobility greatly influence the distribution of this plant.

A. monosperma becomes established in areas where sand movement is slight. It requires light for germination (81) and seeds will not germinate if covered by a thick layer of sand. However, seeds which are completely exposed become

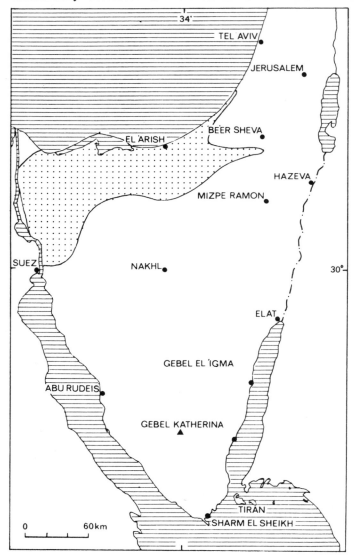

Fig. 88. Distribution map of *Stipagrostis scoparia.*

:::::::: area where the plant is known to occur but distribution pattern is not specified

92

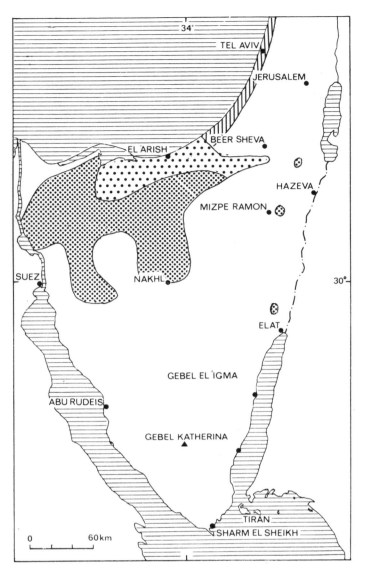

Fig. 90. Distribution map of *Artemisia monosperma*.

area where the plant grows in a diffused pattern

relatively dry area where the plant is restricted to habitats wetter than most of the area (contracted pattern)

relatively wet area where the plant is restricted to habitats drier than most of the area

desiccated and also do not germinate. Suitable conditions for germination occur only when the seeds are covered with approximately 2 mm of sand, and typically occur where there is slight sand movement. Favorable habitats are usually located near existing vegetation less sensitive to sand movement, e.g., *Stipagrostis scoparia* in Districts 6 and 8, and *Ammophila litoralis* in the Mediterranean territory of Israel. After *A. monosperma* becomes established, wind velocity near the plants is reduced, wind-borne silt and clay is deposited, and the water regime gradually improves. Eventually, the perennial grasses that created a suitable habitat for the germination of this wormwood cannot successfully compete, and are eliminated.

It is suggested that the optimal range of *A. monosperma* is the area where the water regime is such that no other species can compete and eliminate *A. monosperma*. In wetter areas, where humus production is substantial, *A. monosperma* cannot compete with more mesophytic species and is replaced in the course of plant succession.

At the arid extremes of its range, this wormwood occurs in relatively moist habitats. Sand fields in western Sinai, which are too dry for *A. monosperma,* are dominated by more drought-resistant psammophytes such as *Cornulaca monacantha* and *Convolvulus lanatus*. The fine-grained sand dunes found scattered throughout these sand fields are sparsely populated by *Stipagrostis scoparia*. At the base of these dunes the long-rooted *A. monosperma* is able to grow, using the rain water that infiltrates through the dune. *A. monosperma* also occurs in sandy wadis in the regs of central Sinai. Here, the dense wadi vegetation reduces wind velocity, causing local sand deposits which provide a suitable habitat for *A. monosperma*. This is clearly seen in the region between Nakhl and Ras el Gindi. The small wadis support communities typical of loess soils with *Artemisia herba-alba* or *Hammada scoparia* as dominants, whereas the larger wadis contain sandy patches with *A. monosperma*.

Growth: In winter new stems with large leaves sprout from buds at the bases of last year's stems. Towards summer, smaller leaves are produced at the distal part of the new stems and in the axils of the winter leaves. Flowering takes place in autumn, by which time the winter leaves are already dry. Achenes are dispersed by wind in winter. Unlike *A. herba-alba*, roots of *A. monosperma* have a thick corky bark that protects from desiccation. This property is important for survival in areas where winds constantly remove sand, exposing more than one meter of the root. Stems covered with a thin layer of sand will develop adventitious roots as well as new stems that grow above the surface. If the sand cover is too deep the plant will die.

Retama raetam (Forssk.) Webb
PAPILIONACEAE

English: white broom.

Hebrew: rotem hamidbar.

Bedouin: ratam, ratim.

Name: The name rotem is mentioned several times in the Bible and in the Talmud. The scientific name is derived from the Bedouin name which in turn is derived from the Hebrew Biblical name.

Chorotype: Saharo-Arabian, extending into the Mediterranean.

Distribution (Fig. 91): White broom is distributed in a diffused pattern as follows: in the sands of the Negev, the Sharon Plain, and Sinai; on weathered sandstone east of the Sea of Galilee and in eastern Samaria; on limestone outcrops in gorges of the Judean and Samarian deserts; on limestone hills covered with mobile sand to depths of 2 m in northwestern

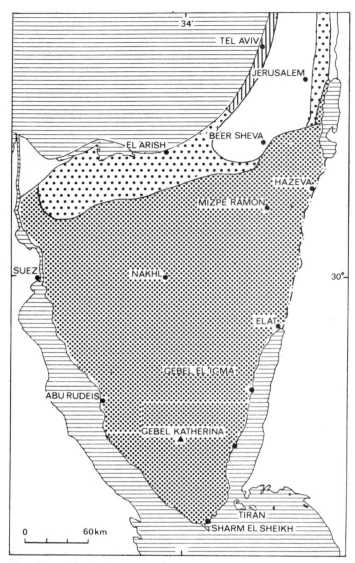

Fig. 91. Distribution map of *Retama raetam*.

area where the plant grows in a diffused pattern

relatively dry area where the plant is restricted to habitats wetter than most of the area (contracted pattern)

relatively wet area where the plant is restricted to habitats drier than most of the area

Sinai (p. 55, Figs. 43 and 44). It grows in wadis in most of the Judean Desert, Negev, and Sinai. The optimal range of white broom is probably the sandy areas of the northern Negev where competing species are absent once the sand becomes stabilized. On the other hand, in the Sharon Plain, white broom gives way to more mesophytic species after the sand becomes stabilized. Therefore, the Sharon Plain may be regarded as wetter than optimal for *R. raetam*. Wadis in the Negev and Sinai are drier than optimal for white broom. In such wadis in Gebel el 'Igma and southern Negev, dead white broom shrubs were observed following successive drought years. However, on rocky substrates in wadis, white broom survives dry years owing to sufficient runoff water even after light showers.

Growth: *Retama raetam* is a stem-assimilant, the stems

photosynthesizing during the dry season. In that period many stems are shed, thereby reducing the transpiring area. Following the winter rains, new twigs bearing small leaves and flowers develop from the green stems. The number of flowers produced depends on the available water. During winter, the new stems are densely hirsute and are sparsely covered with minute scale-like leaves. Towards summer, the leaves fall, the stems harden and most hairs are shed, remaining only in "furrows" where the stomata are located.

The diaspore of *R. raetam* is the entire fruit which contains one to three seeds. The egg-shaped fruits are dispersed by flood water in wadis, by wind rolling them over the sands, and by animals eating the fruits and excreting the seeds. A few varieties of *R. raetam* have been differentiated using seed color and other morphological properties (145).

Uses: The fruits and flowers are eaten by goats and may comprise their only food in certain areas during drought years. Because of its forage value, the Bedouin of southern Sinai prohibit the cutting of white broom stems for fuel or other purposes. The Bible and Talmud mention the high heat value of the charcoal produced from this plant. In central and northern Sinai, at a few sites where it is abundant, the Bedouin use white broom to produce charcoal for sale.

Hammada salicornica (Moq.) Iljin
CHENOPODIACEAE

Hebrew: khammadat haseeakh.

Bedouin: rimth.

Name: The genus name is discussed on page 144. The species is named after *Salicornia* which has similar jointed green stems.

Chorotype: East Sudanian.

Distribution (Fig. 92): *H. salicornica* grows mainly in areas where the sand is derived from the weathering of sandstone. It occurs in wadis and slopes of magmatic and metamorphic rocks and in District 14, on gypsum strata. It is occasionally found on limestone, on alluvium derived from limestone, and in young sands of the Mediterranean coast.

H. salicornica grows in a diffused pattern in sand, and becomes restricted to wadis where the sand is rich in fine-grained particles (Plate 15). Near the Dead Sea it is restricted to sites below sea level. In the Rotem and Yamin plains it is found at elevations of 500 m, whereas in southern Sinai it is found up to 1,300 m above sea level.

Growth: *H. salicornica* is physiologically most active in summer. New stems elongate in spring and summer, flowers appear in autumn, and fruits ripen at the beginning of winter. Following fertilization, the sepals develop into translucent cream-colored wings. Winged diaspores consisting of a one-seeded fruit are dispersed by wind during winter. The green bark on all the stems becomes dry in winter. At the same time the distal parts of the stems that bore flowers and fruits fall. At the end of winter, new stems sprout from buds located at the bases of the previous year's stems. Seedlings have leaves about

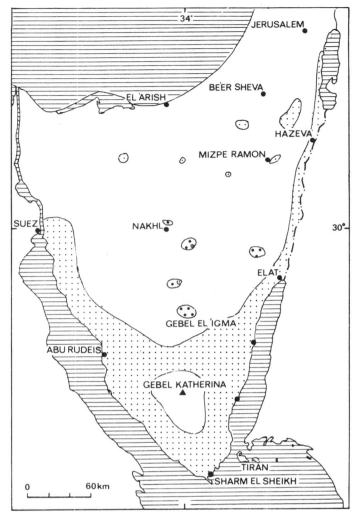

Fig. 92. Distribution map of *Hammada salicornica*.

area where the plant grows in a diffused pattern

area where the plant is known to occur but distribution pattern is not specified

Fig. 93. Galls on *Hammada salicornica* stems. The galls develop after insects lay eggs in the stems. Light gall on left is unopened; dark gall to its right resembles an opened bird's beak.

Plate 15. A stem of *Hammada salicornica* with a drop of sweet fluid excreted by a small insect; the Bedouin in Sinai call such drops "mann el rimth".

l cm long, while in the mature plants the leaves are minute and scale-like. In the axile of these scales are buds covered with hairs. These hairs create a moist microclimate protecting the buds from heat and aridity. *H. salicornica* bears distinctive galls that resemble a bird's beak (Fig. 93).

Uses: Drops of a white sweet fluid probably excreted by insects, are found on vigorous plants during summer (Plate 15). It was once believed that the ancient Israelites used the sweet material from *H. salicornica* when wandering in the desert (20). Bedouin call the fluid from *H. salicornica* "mann rimth" to differentiate it from "mann" or "manna" which is excreted by scale insects living on stems of *Tamarix nilotica* or *T. mannifera*. The long-term and widespread use of the term "manna" by Bedouin has led many scholars to the conclusion that the Biblical manna is the material produced by insects on tamarisk trees.

Moslem scholars who interpreted the Koran 1200 years ago, applied the term "manna" to sweet material found on various plants. Insects living on *Quercus brantii,* an oak growing in Iran and Iraq, excrete such meterial. Inhabitants of the Zagros Mountains call this fluid "mann el samma" meaning "manna of heaven" and use it for preparing candy of the same name.

Haloxylon persicum Bunge
CHENOPODIACEAE

English: saxaul.

Hebrew: prakrak parsee.

Bedouin: ghadha (pronounced radha).

Name : The genus name is derived from the Greek *halos* for salt and *xylon* for tree, describing the saline habitat typical of many species in the genus. *H. persicum* was first described from specimens collected in Persia (Iran).

Chorotype: Irano-Turanian, extending into the Saharo-Arabian.

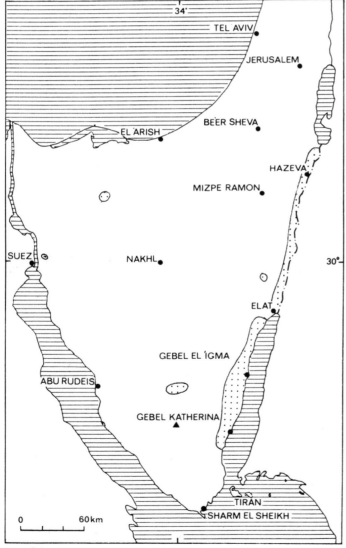

Fig. 94. Distribution map of *Haloxylon persicum.*

:::::::: area where the plant is known to occur but distribution
: : : : : pattern is not specified

Fig. 95. *Haloxylon persicum* growing on sand derived from Lower Cretaceous sandstone (Gebel el Maghara, northern Sinai).

Distribution (Fig. 94): *H. persicum* grows in areas of sand derived from sandstones. It occurs along the 'Arava Valley and in northern and southern Sinai in sand derived from Nubian Sandstone. In Wadi el Haj near the Mitla Pass, where it reaches its westernmost limit, it grows on slopes of sandy alluvium. *H. persicum* shrubs growing on wadi banks are much larger than shrubs growing between wadis. This is probably the result of more available water. The plant often has a huge underground root system.

In Central Asia, *H. persicum* and related species form the "saxaul" woodlands.

Growth : *H. persicum* flowers during February, March, and April. The fruits ripen during the summer and are dispersed in winter.

The plant has galls that resemble flowers with dark petals. It also has globular galls. Neither type of gall is found on any other plant possessing green articulated stems. These galls also distinguish it from *Retama raetam* which it superficially resembles.

PLANTS OF SMOOTH-FACED ROCK OUTCROPS AND CLIFFS

Varthemia iphionoides Boiss. & Bl.
COMPOSITAE

Hebrew: ktela khareefa.

Bedouin: sleemanee, hinedee.

Name : The genus name commemorates L. de Varthemo, a botanist who visited the Near East about 200 years ago. The species name refers to its resemblance (according to Boissier and Blanche) to *Iphiona.*

Chorotype: Mediterranean, extending into the Irano-Turanian and the Saharo-Arabian.

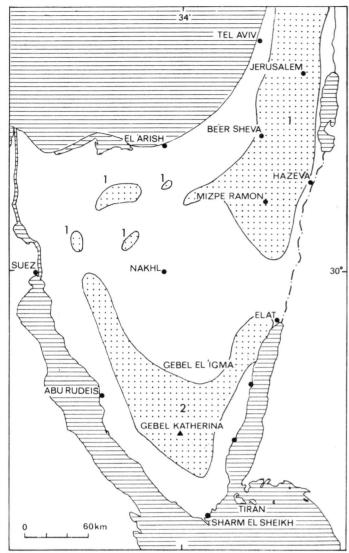

Fig. 96. Distribution map of 1. *Varthemia iphionoides* and 2. *Varthemia montana*.

> [dotted box symbol] area where the plant is known to occur but distribution pattern is not specified

Distribution (Fig. 96): There are two local species of *Varthemia: V. iphionoides* was described from plants collected in the Galilee and Lebanon and *V. montana* from plants of southern Sinai. *V. iphionoides* occurs in the Mediterranean territory of Israel and in desert areas where the mean annual rainfall is above 50 mm. This plant typically grows in the crevices and fissures of smooth-faced limestone and dolomite outcrops (Fig. 97). However, in areas having more than 400 mm annual rainfall it occurs on chalk outcrops as well. Its seeds can germinate in small pits on rock surfaces (150). In the desert it is found mainly in rock outcrops, but is occasionally found in pebbly sections of large wadis and in roadsides. In areas with less than 50 mm of rainfall, there are very few suitable outcrop habitats and *V. iphionoides* occurs mainly in wadis in which the water channel is in hard limestone.

V. montana grows in southern Sinai mainly in sandstone, magmatic, and metamorphic rocks. In central Sinai, there are

Fig. 97. *Varthemia iphionoides* growing in crevices of smooth-faced limestone outcrop.

plants of *Varthemia* that appear to be intermediate between *V. iphionoides* and *V. montana*.

Growth: Following the first winter rains, new branches develop at the base of the stems that bore inflorescences in the previous year (Fig. 98). Winter growth is slow and most stem elongation takes place in spring and early summer. Flowering takes place in September and October and the seeds are dispersed by wind towards winter. After seed dispersal the stems desiccate and very few leaves are present in winter. Leaves and inflorescences are covered with both glandular and non-glandular hairs. The mature leaves have a characteristic strong scent when squeezed.

Uses: In southern Sinai *Varthemia montana* is used by Bedouin for making tea. *V. iphionoides* is said to be too bitter for this purpose.

Fig. 98. *Varthemia iphionoides* in spring. Note the large leaves on new stems and the dry stems from the previous year.

1. *Origanum dayi* Post, 2. *O. ramonense* Danin, 3. *O. isthmicum* Danin

LABIATAE

1. English: Day's origanum, Day's oregano.
1. Hebrew: azoveet hamidbar.
2. Hebrew: azoveet ramon.
3. Hebrew: azoveet Sinai.
1. Bedouin: zu'aytri.
3. Bedouin: za'atar.

Name: The genus name is the common Greek plant name; the first species is named in honor of the botanist Day. The names of the two others refer to the area where their type specimens were collected.

Chorotype: *O. dayi* is endemic to the Irano-Turanian territory of the northeastern Negev and the southern Judean Desert; *O. ramonense* is endemic to the Central Negev Highlands; *O. isthmicum* is endemic to a portion of Gebel Halal in the Isthmic Desert (northern Sinai).

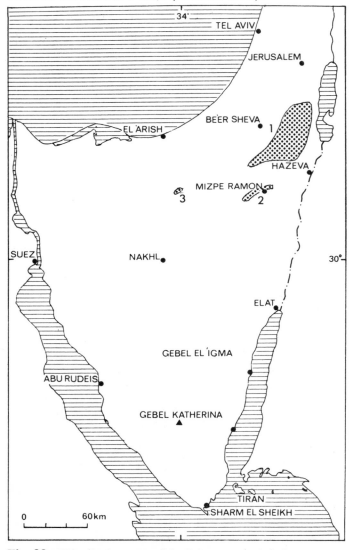

Fig. 99. Distribution map of 1. *Origanum dayi*, 2. *O. ramonense* and 3. *O. isthmicum*.

 relatively dry area where the plant is restricted to habitats wetter than most of the area (contracted pattern)

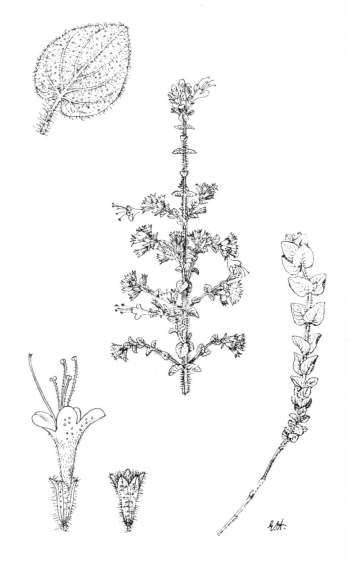

Fig. 100. *Origanum ramonense*, an endemic species of the Central Negev Highlands; flowering stem; a stem in winter; flower (4x); calyx and leaf (4x).

Distribution (Fig. 99): *O. dayi* occurs at elevations from sea level to 600 m in areas having 80 to 300 mm annual rainfall. It is more common in the drier portions of its range (i.e.,80 to 150 mm annual rainfall) than in the moister portions. Its principal habitats are soil pockets of smooth-faced rock outcrops, the foot of such outcrops, and the floor of rocky wadis. In the cooler Central Negev Highlands, *O. dayi* does not grow and its typical habitats are occupied by *O. ramonense*, an endemic with smaller range (Fig. 99). *O. isthmicum*, an endemic of Gebel Halal, also has an extremely limited distribution (Fig. 99). The typical habitats of these three local *Origanum* species are similar, i.e., crevices of smooth-faced limestones and dolomite (16, 17). There is no overlapping in the areas of distribution.

Growth: At the beginning of winter, in all three species new stems develop from buds at the base of last year's stems. The new stems achieve maximum growth during spring. In the axils of the large winter leaves, small branches develop bearing small summer leaves. All three *Origanum* species flower between July and September, by which time the large winter leaves are already shed. A few flowers may be found as late as October and November.

These three species can be easily distinguished not only because they grow in different areas, but also by their morphological characteristics. *Origanum dayi* has a yellow-white corolla and not more than a few hairs on the stems or leaves. The corolla of *O. ramonense* is white-purple when in bud. The entire plant is covered with dense hairs (Fig. 100). *O. isthmicum* has a minute corolla, approximately as long as the calyx. The corolla changes from purple to cream-colored following anthesis. The flowers are solitary in the axils of the bracts (Fig. 48). In the other two local *Origanum* species, corolla is much longer than calyx and there are many flowers in the axils of the bracts.

Uses: *O. dayi* and *O. ramonense* are used as a flavoring or substitute for tea; a few leafy stems are immersed in hot water. Leaves of *O. isthmicum* can be used as a cooking spice in place of commercial "oregano". It has a more delicate taste than all other similarly scented plants of Israel and Sinai, (e.g., *Thymus bovei, Coridothymus capitatus, Majorana syriaca*). All three *Origanum* species can be used fresh or dried.

Centaurea eryngioides Lam.
COMPOSITAE

Hebrew: dardar hakharkhaveena.

Bedouin: digin el badan, leekhiya.

Name: For reasons unknown the genus name is derived from *Kentauros*, the half-man and half-horse of Greek mythology. The species name suggests that it resembles *Eryngium*, at least according to Lamarck. The Hebrew name of the genus appears in the Bible and probably refers to the circles formed by the leaf lobes in the rosette. The Bedouin name means "the ibex' beard" describing the white wool which covers and protects the renewal buds in the leaf axils.

Chorotype: Irano-Turanian.

Distribution (Fig. 101): *C. eryngioides* is widely distributed in hilly areas of the Judean Desert, Negev, and Sinai. It occurs on outcrops of limestone, dolomite, chert, sandstone, granite, and metamorphic rocks. It is often found near waterfalls on outcrops of hard rocks worn smooth by water.

Growth: This plant is a hemicryptophyte and its above-ground parts desiccate in summer. New leaves grow following the first winter showers. In spring a long flowering stalk with several inflorescences develop from each leaf rosette. The flowering head contains tubular florets (Fig. 102 and 103). Each floret has a five-lobed tubular corolla, a tube consisting of five fused anthers, and a style in the center. The anthers release pollen inside the tube (stage 1 in Fig. 103). The growing style with its mass of hairs just beneath its stigma acts as a piston in pushing the released pollen up to the aperture of the anther tube. The stigma has two lobes which come apart only after the style grows and emerges from the anther tube (stage 3

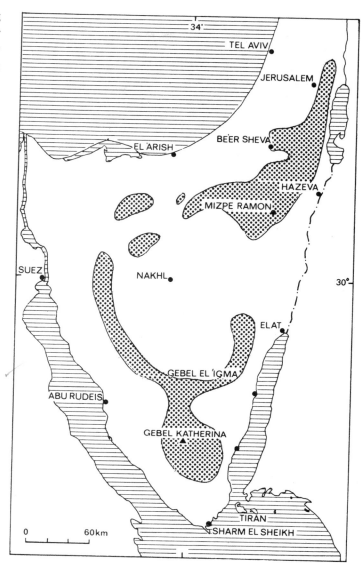

Fig. 101. Distribution map of *Centaurea eryngioides*.

relatively dry area where the plant is restricted to habitats wetter than most of the area (contracted pattern)

Fig. 102. *Centaurea eryngioides*. A flowering head (capitulum) with many florets (see also Fig. 103).

99

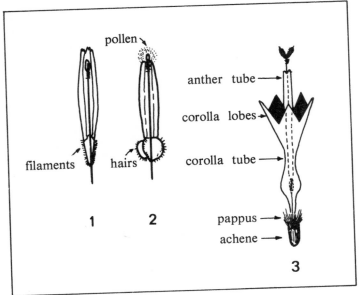

Fig. 103. *Centaurea.* Stages in floret development. 1. A closed stigma within the tube of fused anthers. 2. Stigma emerging from the anther tube. Filaments curved as a result of contact.3. A complete mature floret showing the elongated style with two-lobed stigma.

in Fig. 103). This process takes place without the intervention of insects. However, at stage 1 (Fig. 103), if the anther tube is touched by a nectar-seeking insect, the tube is suddenly pulled down by the contracting filaments, and pollen is deposited on the insect (stage 2 in Fig. 103). Cross pollination occurs when the insect alights on another inflorescence in which the florets have opened stigmas.

Careful removal of the corolla tube reveals five stamen filaments with their sensitive hairs (Fig. 103). The contracting action is a result of a sudden decrease in the turgor of filament cells facing the style. Filaments may remain contracted for at least half an hour following contact. The curved filaments straighten when their cells become turgid again.

Following the ripening and dispersal of achenes in early summer, the above-ground parts of the plant dry out. In the leaf axils, hairs cover the renewal buds and protect them from desiccation during the summer.

1. *Sternbergia clusiana* (Ker-Gawler) Spreng.
2. *Narcissus tazetta* L.
AMARYLLIDACEAE

1. Hebrew: khelmoneet gdolah.
2. English: narcissus; daffodil,
2. Hebrew: narkees matsuy.

Name: *S. clusiana* is named in honor of the botanists Sternberg and Clusius. The Hebrew name refers to the yellow corolla of the flower (Plate 4) whose color is similar to egg yolk (khelmon in Hebrew).

In Greek mythology *Narcissus* was a youth who became so entranced with his reflection that the gods turned him into a flower (120).

Chorotype: Mediterranean.

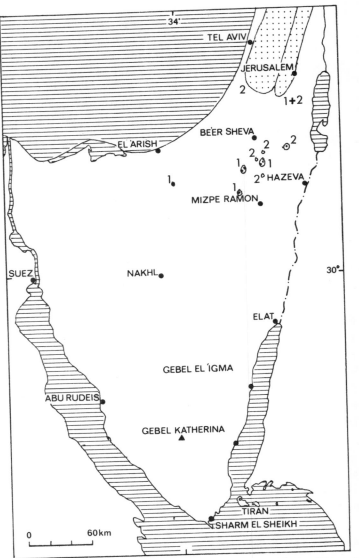

Fig. 104. Distribution map of *Sternbergia clusiana* (1) and *Narcissus tazetta* (2).

[::::::] area where the plant is known to occur but distribution pattern is not specified

Distribution (Fig. 104): Both are Mediterranean species that are represented in the desert only by several isolated populations. *N. tazetta* has been found at four locations in the Negev Highlands and *S. clusiana* at three. These geophytes inhabit crevices of smooth-faced limestone and dolomite outcrops, soil at the foot of these outcrops, and wadis receiving runoff water from such outcrops.

The Negev populations of these species are separated from the Mediterranean populations and from each other by distances much greater than can be accounted for by normal diaspore dispersal. A distance of 50 to 60 km separates the central and northern Negev populations of *S. clusiana*. The distance between the northern Negev and southern Judean Mountains populations is 60 km. The distance between the four Negev populations of *N. tazetta* is 10 to 20 km, and from these to the nearest Mediterranean population the distance is 40 to 50 km. The occurrence of these species isolated in the desert can hardly be regarded as a case of long-distance dispersal. Seeds of *S. clusiana* are dispersed by ants which consume only the oil-body on the seed, leaving the rest of the seed intact. The seeds of *N. tazetta* fall near the mother plant.

S. clusiana and *N.tazetta* are two of nearly 100 Mediterranean species found in smooth-faced rock outcrops of the Negev and Sinai. It has been suggested that these plants thrived in this region thousands of years ago during a period of relatively moist climate (19). As discussed on page 43, the climate of the Middle East and North Africa during the Late Pleistocene was wetter than at present. Pollen of oak, pistachio, olive, and pine found in excavation sites in the Central Negev Highlands was determined to be 22,000 to 50,000 years old (66). In such periods the Mediterranean region extended deep into the area which is now desert. This enabled species such as *N. tazetta* and *S. clusiana* to penetrate this area. When the climate became arid, these species survived as relict populations at sites having a Mediterranean-like moisture regime.

Growth: *S. clusiana* flowers in the Negev in October and November before the development of any leaves. Plants on north-facing slopes bloom first, followed by those on south-facing slopes which flower at the end of October and the beginning of November. Finally, the plants in the wadis bloom in November. Leaves sprout following the first rains, at which time the flower stalk elongates. The fruits mature in winter. The black seeds are partly surrounded by white tissue having a high oil content. *S. clusiana* also reproduces vegetatively. The large bulbs split and each segment gives rise to a new plant.

N. tazetta starts flowering in December, about two weeks after the first rains effectively wet the soil pockets. In each population flowering continues for a month or two. Following effective rains, inflorescences each with two leaves, appear very rapidly, suggesting that most of the cell divisions involved in the initiation of flower parts takes place long before the actual blossoming (Fig. 105) The effective rain merely serves as a trigger for their final development which mostly involves cell elongation. The seeds ripen two months after the flowers bloom. When the tall peduncle is shaken by wind, the dehiscent fruit releases its seeds near the mother plant.

Primula boveana Decne.
PRIMULACEAE

English: Sinai primrose.
Hebrew: bekhor aveev Sinaee.

Name: The genus name, derived from the Latin *primus* meaning first, refers to its early flowering. The species is named in honor of Bové who collected the type specimen in 1832.

Chorotype: Irano-Turanian. Endemic to Sinai.

Distribution (Fig. 106): This is one of the rarest endemic species of Sinai. It occurs at elevations above 1,700 m near small springs which flow throughout the year. These springs are found mostly on red granite. Sites supporting the Sinai

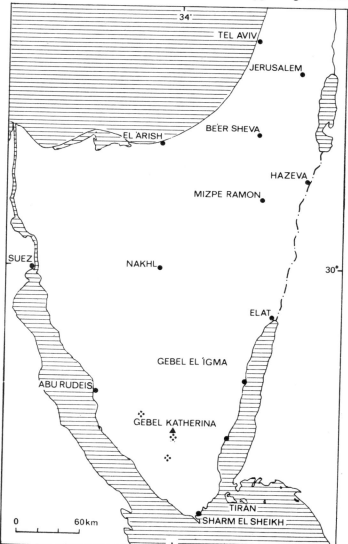

Fig. 106. Distribution map of *Primula boveana*.

Fig. 105. *Narcissus tazetta* in bloom, growing here in rock crevice southwest of Dimona (northern Negev).

relatively dry area where the plant is restricted to habitats wetter than most of the area (contracted pattern)

primrose can often be located from a distance by the dark strips of algae which grow in this habitat. These dark strips of algae alternate with white strips of calcium carbonate deposited as a result of algal activity and the evaporation of water.

Four closely related species are local endemics; one in Yemen and the Horn of Africa, one in Turkey, and two in the Himalayas. Wendelbo (130) suggests that these five endemic species are all derived from one ancestral species that was extensively distributed in Asia and Africa during a cool wet period about six million years ago. With the advent of a drier and warmer climate, several isolated populations survived near mountain springs where they eventually evolved as separate species.

The Sinai primrose has thus far been found on Gebel Katherina, Gebel Musa, Gebel Safsafa, Gebel Serbal, and Gebel Umm Shaumar, mainly on north-facing cliffs.

G r o w t h : Leaf rosettes survive throughout the year. Flowering takes place from February to May, occasionally even in summer (Pl. 14). Minute seeds like those produced by *Primula* are often referred to as "dust seeds". This species, like many other plants growing in moist habitats, produces millions of seeds carried by wind. Although most of the seeds fail to germinate, the large number ensures that at least some will land in the appropriate moist habitat.

Pistacia atlantica Desf.
ANACARDIACEAE

English: atlantic pistachio.

Hebrew: elah atlanteet.

Bedouin: butum.

N a m e : The genus name is derived from the old Persian name "pisté", for the tree bearing pistachio nuts (*Pistacia vera*). The species is named after the Atlas Mountains in North Africa where Desfontaine had collected the type specimen. The name *elah* occurs several times in the Bible.

C h o r o t y p e : Irano-Turanian.

D i s t r i b u t i o n (Figs. 27, 107): Dense stands of *P. atlantica* trees occur in the northeast Galilee between Safad and Kiryat Shmona (mean annual rainfall 600 to 700 mm). Several dozen trees grow near Jerusalem and Hebron, in an area having 500 mm of rainfall. About 1,400 trees are known in the Negev Highlands at elevations of 600 to 1,000 m (29). The greatest density in the Negev is 20 to 30 trees per square kilometer near Har Romem and Har Loz. Suitable habitats are: soil pockets in smooth-faced rock outcrops, soil at the foot of such outcrops, and wadis.

P. atlantica specimens vary greatly in form, depending on the moisture regime at the specific site. It may range from a small non-flowering shrub to a large flowering tree. This is also true of many other tree species growing in deserts. In order to explain the form, size, and density of trees in deserts, we shall discuss in detail how moisture regime affects *P. atlantica*.

The water available to each plant depends on the mass of substrate associated with the roots (rhizosphere), the size of the

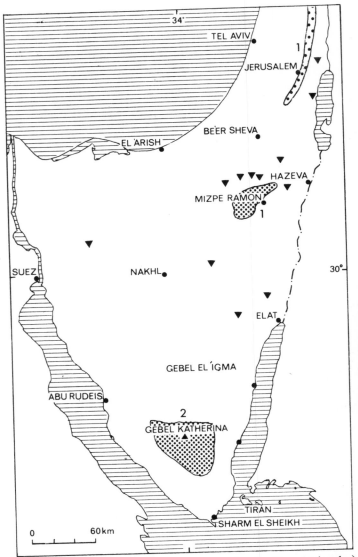

Fig. 107. Distribution map of: *Pistacia atlantica* (1 and triangles), *P. khinjuk* (2).

area where the plant grows in a diffused pattern

relatively dry area where the plant is restricted to habitats wetter than most of the area (contracted pattern)

▼ one or a few individuals

catchment area, the annual rainfall, and the amount of water which eventually reaches the rhizosphere. This last factor depends on the properties of the catchment area that contributes runoff water and on the number of days in which there is runoff. Smooth-faced rock outcrops contribute much more runoff than most other substrates. On smooth rocks runoff may start after the first millimeter of rain. The running water first saturates the soil pockets in the rocks and then, if the rain continues, the foot of the rock outcrop is watered. After strong showers the runoff may reach the wadis. When compared to smooth rocks, the "yields" of runoff from stony slopes or slopes of highly fissured rocks is much lower, and heavier showers are needed to create runoff. In the Negev Highlands small catchment areas of stony slopes contribute

Fig. 108. *Pistacia atlantica* in Nahal **Eliav, Central** Negev Highlands.

runoff only after at least 4 to 6 mm of rain falls within a short period of time.

Large catchment areas contribute runoff only after strong rains of 10 to 15 mm (115). Light showers are much more common than strong showers (116). Thus, in stony slopes, the larger the catchment area the fewer the number of days in which there is runoff. The following habitats of *P. atlantica* in the Negev are listed in order of decreasing frequency of days in which rain causes runoff: 1. crevices in smooth-faced rock outcrops 2. foot of rock outcrop 3. small wadis with smooth rocks in their catchment area 4. small wadis with stony soil or fissured rocks in their catchment area 5. large wadis.

The rhizoshere appears to be smallest in rock crevices, larger at the foot of rock outcrops, and most extensive in large wadis where the ground consists of gravelly alluvium (123). The large rhizosphere may explain the developement in the wadis of the Central Negev Highlands of *P. atlantica* trees as tall as 15 m and having a trunk 2 to 3 m in diameter (Fig. 108). At the foot of rock outcrops there are trees 4 to 5 m high with trunk diameter of 0.5 to 1 m (Fig. 109). Rock crevices may support small shrubs of *P. atlantica* up to 150 years old. The shrubs here are limited in their size by the small space available for their rhizosphere.

The high density of *P. atlantica* trees near Har Romem and Har Loz (Fig. 27) is related to the existence there of smooth-faced rock outcrops. Many trees occur here at the foot of these outcrops and in wadis draining slopes with these rocks. Based

on aerial photographs, the size of the catchment area of individual *P. atlantica* trees was measured (29). When measuring the catchment areas, the only tree studied in a given wadi or slope was the one closest to the ridge. It was found that the mean area for nine catchments with stony slopes was 250,000 square meters, and that the mean area for seven catchments with smooth rocks was only 1,500 square meters. This illustrates the superiority of smooth rocks as a habitat for trees in the desert. The solitary *P. atlantica* trees near Be'er Milhan, Gebel Shaira, and Gebel Sahaba are restricted to wadis. Here the slopes consist of fissured limestone and stony soil.

Fig. 109. *Pistacia atlantica* 4 to 5 m tall at the foot of smooth-faced rock **outcrops, Central** Negev Highlands.

Many examples of trees growing in association with smooth rocks are found in deserts throughout the world. In the Negev the following trees and shrubs are found mostly on smooth rocks: *Ceratonia siliqua*, *Amygdalus ramonensis*, *Rhus tripartita*, and *Rhamnus dispermus*. In Sinai this habitat supports: *Crataegus sinaicus*, *Pistacia khinjuk*, *Cotoneaster orbicularis*, *Sageretia brandrethiana*, *Juniperus phoenicea*, *Rhamnus dispermus*, and *Rhus tripartita*. In Jebel Ahagar (Algerian Sahara) *Cupressus dupreziana* and *Olea laperrinii* occur in such rocks (107). In the desert areas of Joshua Tree National Monument (California, U.S.A.), several species of *Juniperus*, *Pinus*, and *Quercus* thrive in fissures and crevices of smooth rocks which are very similar to those of southern Sinai.

P. atlantica is absent in southern Sinai. However, a related species, *P. khinjuk*, is found here. It is distributed mainly at elevations above 900 m, and is most abundant in Gebel Serbal, at elevations of 1,700 to 2,000 m. This mountain of smooth-faced granite (Pl. 2) supports hundreds of trees in habitats analogous to those of *P. atlantica* in the Negev Highlands. Outside of Sinai, populations of *P. khinjuk* are found in Gebel Galala in Egypt and probably also in Petra, Jordan. The major distribution area of this species is in Iran and Afghanistan, where "khinjuk" is the vernacular name.

Growth: Seedlings one, two, and three years old of *P. atlantica* and *P. khinjuk* have been found in the relatively moist habitats of the desert. The greater number of seedlings is found in rock crevices; there are fewer at the foot of smooth rock outcrops and fewest in wadis. Whereas the moisture in crevices is favorable for germination, the rhizosphere there is too small to support large flowering trees.

Like all species of *Pistacia*, *P. atlantica* is dioecious and wind pollinated. Flowers open in spring and leaves sprout almost at the same time. The shrubs growing in rock crevices do not flower and start shedding their leaves in August. The flowering trees in wadis and at the foot of rock outcrops shed their leaves in October. Those fruits containing one seed turn bluish-green as they ripen in autumn. Fruits without seeds grow to the same size as full fruits but are red. Fruits are dispersed by flowing water in wadis and most probably by birds which eat the fruit and excrete the undigested shell containing the seed.

Uses: The ripe fruit can be eaten either fried or raw after removing the shell. The *P. atlantica* tree is used as stock onto which are grafted scions of the commercial pistachio nut *P. vera*. Using such grafts, several pistachio nut plantations have already been established in the Negev.

Juniperus phoenicea L.
CUPRESSACEAE

English: Phoenician juniper.

Hebrew: 'ar'ar adom.

Bedouin: 'ar'ar.

Chorotype: Mediterranean.

Distribution (Figs. 110 and 111): *J. phoenicea* occurs throughout the Mediterranean coastal region except for the deserts of Libya and Egypt. In northern Sinai the species is mostly confined to areas containing smooth-faced limestone outcrops; thousands in Gebel Halal, hundreds in Gebel Maghara, and dozens in Gebel Yiallaq (Fig. 46).

In common with *Pistacia atlantica* it occupies rock outcrops, the base of these outcrops, and wadis. In Petra, Jordan (ca. 100 km northeast of Elat) it occupies similar habitats in outcrops of massive sandstone. North of Petra, *J. phoenicea* together with *Quercus calliprinos* form a maquis (146).

Palaeobotanic records of *J. phoenicea* date from 30,000 to 40,000 years ago in the Negev near Urim (J. Prior, personal communication), from 4,000 to 34,000 years ago in Gebel Maghara (60), and from 10,000 years ago in Har Harif and Har Horsha southwest of Mizpe Ramon (66). These findings indicate that *J. phoenicea* was once more widespread in the

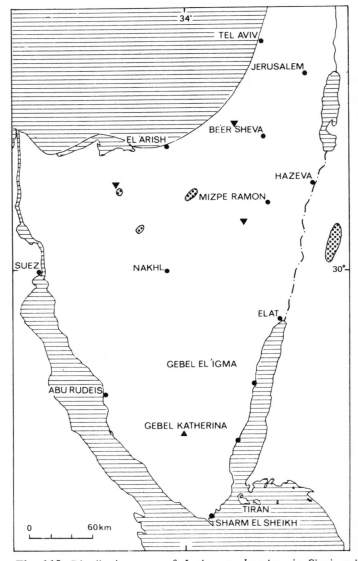

Fig. 110. Distribution map of *Juniperus phoenicea* in Sinai and Israel. ▼ indicates a fossil record.

relatively dry area where the plant is restricted to habitats wetter than most of the area (contracted pattern)

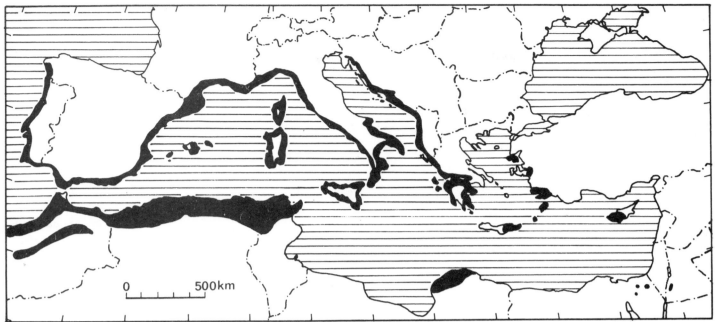

Fig. 111. Distribution map of *Juniperus phoenicea* in the Mediterranean basin.

Negev and Sinai and once grew in the Negev Highlands, where it is now extinct.

In Greece, *J. phoenicea* is a dominant in the maquis west of Delfi. In this maquis, which receives an annual precipitation of 500 to 600 mm, *J. phoenicea* is accompanied by *Olea europaea*. This is probably part of the optimal range for *J. phoenicea*. In southern France and eastern Pyrenees, where the annual rainfall is 900 to 1,000 mm, it is confined to chalky marl outcrops and to cliffs of hard limestone. The small size of the soil pockets in crevices and fissures of the limestone cliffs restricts the amount of water available to plants. *J. phoenicea*, which is more drought-resistant than most maquis components of the wet area, successfully competes with these mesophytes in the limestone cliffs.

Growth: The xylem of this tree is rich in resins which protect it from insects and fungi. Trunks of dead trees 400 to 600 years old are common at Gebel Halal. One living branch of a tree inhabiting a rock outcrop was found to have 860 annual rings, varying considerably in width. A dendrochronological study of *J. phoenicea* from Gebel Maghara and Gebel Yiallaq indicated that annual rainfall varied from periods with 30 mm to periods with 300 mm of annual rainfall. The rainfall in this area now averages 100 mm. (127). *J. phoenicea* resembles *Zygophyllum dumosum* (Fig. 20) in that individual sections of the trunk are active as long as the connected root to that section supplies it with water. By the end of 1978, for example, many juniper trees in the wadis of Gebel Halal had become partially desiccated, suggesting that there were several consecutive dry years. The remaining branches on the same trees were still completely vital.

Year-old seedlings are found in rock crevices, in rocky wadis, and at the foot of rock outcrops. In three different years such seedlings were found in Gebel Halal and Gebel Maghara.

Dead trees are found in Gebel Halal scattered among the living trees. The dead trees are found in somewhat marginal microhabitats such as at the foot of small limestone outcrops. Here the *J. phoenicea* trees die after several successive dry years.

Use: Some Bedouin say they do not use this tree for fuel because the sparks produced by the burning stems burn their tents. However, they do use the resin to produce incense for religious purposes.

Rhus tripartita (Ucria) Grande
ANACARDIACEAE
English: Syrian sumac.
Hebrew: og kotsansee.
Bedouin: 'irn.
Name: The genus name is derived from the common name in Greek for *Rhus coriaria*. The species name describes the three leaflets of each leaf (Fig. 112).

Fig. 112. *Rhus tripartita*

Chorotype: West Irano-Turanian, extending into the Mediterranean and the Saharo-Arabian.

Distribution (Fig. 113): *R. tripartita* grows in the deserts of Israel and Sinai in areas with cliffs and canyons. It is found in deep-fissured limestone, dolomite, magmatic, and metamorphic rocks all of which weather into large blocks. In the deserts of Israel and Sinai the greatest density of *R. tripartita* is found in canyons in the northern Judean Desert and eastern Samaria. It grows also in the Mediterranean territory east of Kefar Sava as a component of the *Pistacia lentiscus – Ceratonia siliqua* association.

It is likely that this species was more widely distributed when the Negev and Sinai had a moister climate. This conclusion is based on the facts that today *R. tripartita* grows in dense stands in the Mediterranean territory, while it is restricted to relatively wet microhabitats in the desert.

Growth: *R. tripartita* is a shrub 1 to 3 m high. In summer it gradually sheds its leaves. New leaves and flowers cover the stems following the first winter rains. Red berries develop in spring and are probably bird-dispersed.

Uses: The Bedouin of Sinai boil the young lignified stems to prepare a drink which resembles tea in color and taste. The fruits are not edible.

COLONIZING SPECIES
Atriplex leucoclada Boiss.
CHENOPODIACEAE
English: white-branched orache.
Hebrew: maluakh malbeen.
Bedouin: rughul.
Name: The genus name is derived from the common name in Greek of another *Atriplex* species. The species name means white-branched. The genus name in Hebrew means salty and refers to the salty taste of its leaves. "Malbeen" means white.
Chorotype: Saharo-Arabian and Irano-Turanian.

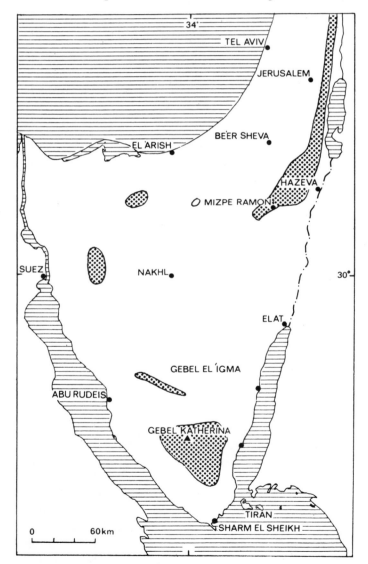

Fig. 113. Distribution map of *Rhus tripartita*.

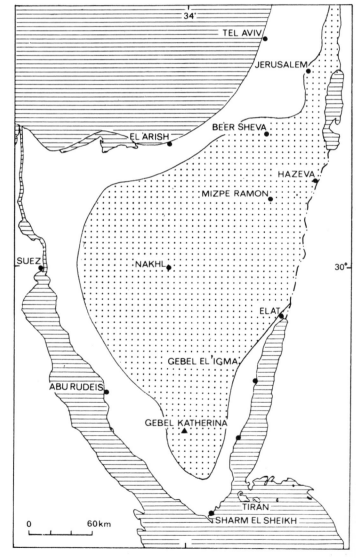

Fig. 114. Distribution map of *Atriplex leucoclada*.

relatively dry area where the plant is restricted to habi–tats wetter than most of the area (contracted pattern)

area where the plant is known to occur but distribution pattern is not specified

Distribution (Fig. 114): *A. leucoclada* grows mainly in recently disturbed habitats in which the vegetation has been destroyed. Among these disturbed habitats are wadi beds, where floods periodically remove the top soil including the seeds. When wind-borne *A. leucoclada* diaspores land here, they can establish themselves without competition. Similarly, it inhabits roadsides and construction areas where the topsoil is removed.

It also grows in sites where the existing vegetation has been destroyed by drought. At Gebel el 'Igma, *Salsola tetrandra* once dominated all the slopes at elevations of 1,000 to 1,200 m. After several consecutive drought years, all the *S. tetrandra* shrubs died. After the rains of 1968, *A. leucoclada* plants became established on these slopes. At elevations of 1,500 to 1,600 m at Gebel el 'Igma, *Artemisia herba-alba* once dominated the northern slopes, but died after the drought years. Here again, *A. leucoclada* became established after the rains of 1968. On the southern slopes at the same elevations, *Halogeton alopecuroides,* more drought resistant than *A. herba-alba,* did not die, and *A. leucoclada* was not able to become established after the rains of 1968. A few years after *A. leucoclada* colonizes an area, diaspores of species better adapted to that environment may arrive and take over.

Growth: New stems develop in winter from buds near the base of older stems. They elongate in spring and summer. Flowering takes place from September to November, and fruits ripen at the beginning of winter. Unlike many other members of the Chenopodiaceae, *Atriplex* diaspores consist of fruits not accompanied by sepals but by two triangular or deltoid bracts. The diaspores of *A. leucoclada* fall near the mother plant and may be carried long distances by flood water or strong winds. Plants that germinate in winter may bear flowers the following autumn.

A. leucoclada is a perennial semishrub in those habitats where water supply is sufficient and an annual in drier sites. Individuals within a population exhibit a great diversity of morphological characters. They may vary with respect to shape of leaves, morphology of diaspores, hair density, and duration of flowering. Individual stems may carry up to five different forms of diaspores. In *Atriplex rosea* and several Australian *Atriplex* species, it was found that diaspores which differ morphologically also differ in their germination properties (69). The same may be true for *A. leucoclada*.

Uses: The leaves of *A. leucoclada* can be eaten fresh or cooked, but are not as tasty as those of *A. halimus.*

Anabasis setifera Moq.
CHENOPODIACEAE
Hebrew: yafruk zifanee.
Bedouin: gilu, gilwe.
Name: The genus name is discussed on page 77. The species name refers to the leaf bristle, prominent in some populations but mostly absent from this species in Israel and Sinai.

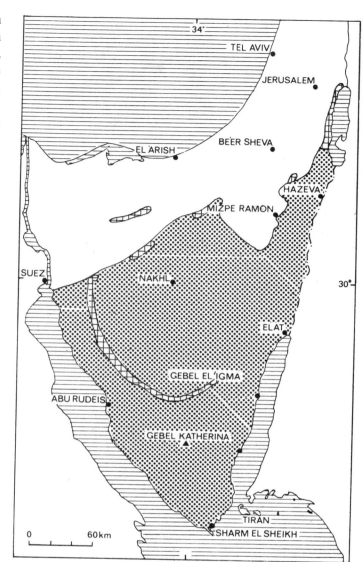

Fig. 115. Distribution map of *Anabasis setifera.*

area with optimal water regime (diffused pattern on various soil types and slopes)

relatively dry area where the plant is restricted to habitats wetter than most of the area (contracted pattern)

Chorotype: East Saharo-Arabian, extending into the Irano-Turanian.

Distribution (Fig. 115): *A. setifera* is dominant on newly exposed pebble and gravel substrate along the receding Dead Sea shoreline. It grows as a dominant in escarpments of large geomorphological structures. These include the fault escarpment near the Dead Sea, folding escarpments of anticlines in the Negev and Sinai, and the erosion escarpments of Gebel et Tih in Sinai. In these escarpments it occupies colluvium that is actively shifting downslope. It is also a dominant near newly-constructed roads built on such escarpments, as well as in wadis. Other habitats include salt marshes in the 'Arava Valley and wadis having chalky ground in central Sinai.

Anabasis setifera is a colonizer that lives longer and occupies warmer areas than *Atriplex leucoclada.*

Fig. 116. Succulent stems of *Anabasis setifera* with a drop of sweet liquid. A few ants attempting to feed on the liquid were trapped.

Growth: *A. setifera* begins intensive stem growth in summer. These elongating stems are bright green, in marked contrast to the dry desert slopes. Flowering takes place at the end of the summer, and the fruits together with the winged sepals are dispersed by winter winds. In winter the distal portion of each branch that bore fruits becomes dry, and in the following summer new stems develop from buds at the base. Each plant produces a large number of winged diaspores, which facilitates the rapid establishment of the species in available habitats.

A. setifera is a succulent which is physiologically active in summer. Several insects lay their eggs in the stems stimulating the formation of galls of various types. One kind of gall is a swollen portion of the stem; another resembles a ball covered with minute hairy leaves. In the Dead Sea Valley insects penetrate new internodes during August and September, causing formation of sweet white liquid drops similar to the "manna" of *Hammada salicornica* (Fig. 116).

Halogeton alopecuroides (Del.) Moq.
CHENOPODIACEAE
Hebrew: mlekhaneet ha'aravot.

Bedouin: sh'aran.

Name: The genus name refers to the plant's affinity for salty soils. The species was named by Delile and refers either to the genus *Alopecurus* (Graminae) or to a fox's tail.

Chorotype: Saharo-Arabian.

Distribution (Fig. 117): *H. alopecuroides* grows as a dominant in a diffused pattern on soils derived from limestone, chalk, and marl in the Negev, Judean Desert, and central Sinai. In southern Sinai it is confined to dark magmatic rocks. It is also common as a dominant in wadis where the ground is chalky. Its habitat is typically drier than that of *Artemisia herba-alba* and moister than that of *Zygophyllum dumosum*. A common associate is *Anabasis articulata*. In Gebel el 'Igma, at

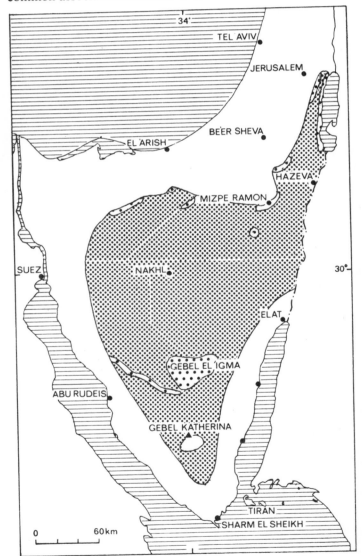

Fig. 117. Distribution map of *Halogeton alopecuroides*.

area where the plant grows in a diffused pattern

relatively dry area where the plant is restricted to habitats wetter than most of the area (contracted pattern)

108

elevations of 1,200 to 1,400 m, it grows in association with *Salsola tetrandra* on all slopes. At higher elevations *H. alopecuroides* dominates the southern slopes. In many areas it is the dominant of wadis having a chalky substrate.

Another important habitat consists of unstable soil on escarpments as described for *Anabasis setifera* and *Atriplex leucoclada* (see above). However, *H. alopecuroides* is mainly a species of stable habitats and is a colonizer only in small areas. On the other hand *Atriplex leucoclada* is mainly a colonizer and grows in undisturbed habitats only in a few areas. *Anabasis setifera* is an intermediate between the two.

Growth: *Halogeton alopecuroides* may withstand several drought years by drastically reducing its transpiring area. The leaves then resemble small balls. In the spring of rainy years, new stems develop from buds near the base of last year's branches. The new stems bear cylindrical leaves, 1 cm long, most of which terminate in a bristle. Flowering takes place in autumn. Winged diaspores resembling those of *Anabasis setifera* are dispersed by winter wind.

Uses: This is an important forage plant for goats and camels in summer. At Gebel el 'Igma, it provides a dependable supply of forage even in drought years. The Bedouin say that eating the sour-tasting *H. alopecuroides* (as well as several other members of the Chenopodiaceae) helps rid the goats of the intestinal illness they acquire by eating green grass. Possibly, the intestinal illness is caused by parasites that flourish on grass in rainy years (5).

Pulicaria desertorum DC.
COMPOSITAE

English: desert fleabane.

Hebrew: par'osheet galoneet.

Bedouin: rabl, rabil.

Name: The genus name refers to the supposed effectiveness of the plant's scent in repelling fleas, according to ancient European folklore. *Pulex* means flea in Latin. The Hebrew genus name also refers to fleas. The Bedouin of the Negev and Sinai do not ascribe such flea-repelling properties to this plant.

Chorotype: East Saharo-Arabian and east Sudanian.

Distribution (Fig. 118): *P. desertorum* is a plant of warm desert areas and is not found in the relatively cool parts of the Judean Desert, northern Negev, and northern Sinai. It colonizes barren areas where the salts are completely leached out. This is in contrast to the three previously discussed colonizers which inhabit somewhat saline soils.

Successive dry years which lead to the desiccation of the existing vegetation, or floods which remove the topsoil with the existing plants and seeds, are the two most important habitats available for the colonization of *P. desertorum*. In most of its range it is restricted to wadis. However, in southern Sinai the dark-colored rocks absorb enough heat and receive enough precipitation to support the growth of *P. desertorum* in a diffused pattern. Following effective winter rains, the plant may cover extensive areas, such as the granitic alluvial fans north of Sharm el Sheikh. In these fans plants dry after one or

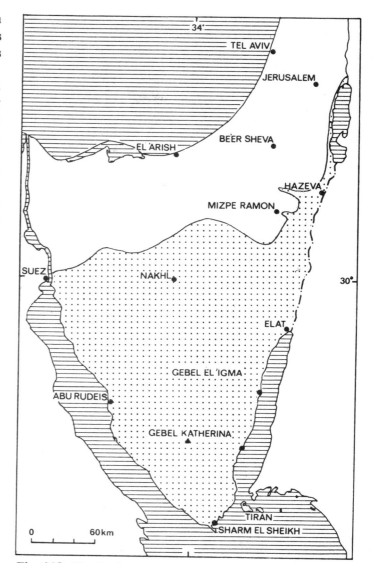

Fig. 118. Distribution map of *Pulicaria desertorum*.

area where the plant is known to occur but distribution pattern is not specified

two years without rain; in rocky habitats where moisture is concentrated in fissures, they can survive a longer drought.

The biomass of *P. desertorum* may vary considerably from year to year. For example, the granitic alluvial fans north of Sharm el Sheikh were extensively covered with *P. desertorum* in a rainy year; they only supported a single line of plants along the wadis in subsequent dry years.

Growth: *P. desertorum* is either an annual or perennial, depending on moisture conditions. Germination takes place in winter or spring. Plants that germinate in late spring are more densely covered with hairs than plants that germinate in winter. The green leaves and stems have a distinctive pleasant scent somewhat resembling that of peaches.

In spring the plant produces yellow inflorescences about 1 cm in diameter, the number of which depends on the available moisture. Towards summer, each plant produces thousands of minute wind-borne achenes that enable the plant to rapidly occupy suitable sites.

Uses: A very tasty drink is made by immersing a few leafy stems in boiling water or by adding a few leaves to tea. Leaves dried in spring retain their good taste for a long time.

ANNUAL SPECIES

Stipa capensis Thunb.
GRAMINAE

English: twisted-awn feather grass, Cape feather grass.

Hebrew: mal'aniel matsuy.

Bedouin: safsuf.

Name: The genus name is based on the common Greek name of another species in the genus. The Hebrew name refers to its prominent awns. The common name, feather grass, refers to the feather-like awns of many species in this genus. However, the awns of *S. capensis* are not feather-like. One of the synonyms for this plant is *S. tortilis* Desf., where *tortilis* means twisted in Latin and refers to the twisted awns.

Chorotype: Irano-Turanian and Saharo-Arabian, extending into the Mediterranean.

Distribution: This is one of the most common annuals of Israel and Sinai. It grows in areas where the mean annual rainfall ranges from 20 mm to 700 mm. Dense stands occur on plains and slopes in the northern Negev. In moister areas it is restricted to southern slopes having shallow soil. In more arid sites it is the dominant annual on rocky habitats covered by *Zygophyllum dumosum* associations. Sparse populations of *S. capensis* are found in small wadis.

Growth: *S. capensis* germinates at the beginning of winter; the seeds ripen in spring and are dispersed at the beginning of summer. The seeds, armed with a twisted awn and a sharp base, become attached to the hair of animals or are dispersed by wind. They are also capable of "creeping" on soil by the uncoiling and recoiling of the twisted awn as a result of changes in air humidity. When obstructed by stones or dead plants, instead of creeping, the diaspore bores into the ground by means of the circular movement of the sharp pointed base.

Studies of competition between *S. capensis* and *Avena sterilis,* one of the most common annual grasses of the Mediterranean habitats in Israel showed that when moisture is not limiting, *A. sterilis* takes over the area (122). However, drier habitats are dominated by *S. capensis,* and *A. sterilis* is restricted there to wetter microhabitats.

Mesebryanthemum nodiflorum L.
AIZOACEAE

English: fig marigold.

Hebrew: ahal matsuy.

Bedouin: shnin.

Name: The genus was originally named *Mesembrianthemum* by Breyne in 1684, based on the Greek *mesembria* for noon and *anthemon* for flower. He was only familiar with a species which flowered at noon. The genus now includes some species that flower at night. In 1719 Dillenius proposed that *i* be changed to *y,* and the present name is based on the Greek

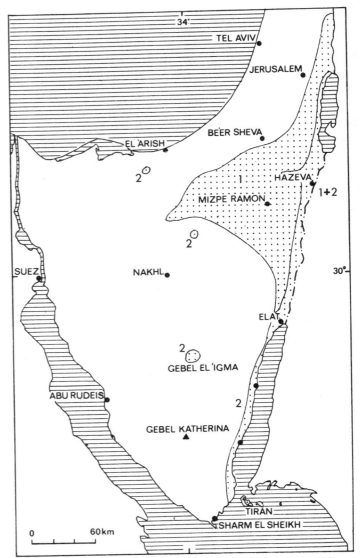

Fig. 119. Distribution map of (1) *Mesembryanthemum nodiflorum* and (2) *M. forsskalii,* the area in which the ranges of the two species overlap is indicated by 1+2.

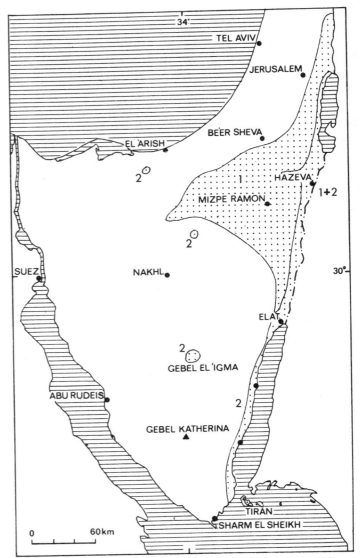 area where the plant is known to occur but distribution pattern is not specified

mesos for middle, *embryon* for embryo, and *anthemon* for flower.

Chorotype: Mediterranean, Euro-Siberian, and Saharo-Arabian.

Distribution (Fig. 119): *M. nodiflorum* is a halophyte growing mainly in the Dead Sea Valley, the southern Negev, and the 'Arava Valley. In studies of several halophytes in Israel, it was found that for a given combination of soil and climate, one species is better adapted than any other (23, 27). *M. nodiflorum* is a dominant in associations of annual species, and in highly saline sites it often forms monospecific stands. *M. nodiflorum* is most successful in soils which are saline from the surface down to at least 20 cm (23). In the area where *M. nodiflorum* is dominant, small depressions leached by large quantities of water support non-halophytes such as *Trigonella stellata, Stipa capensis, Plantago ovata,* and many others.

Because of competition at these leached sites, *M. nodiflorum* plants remain small and never flower.

The fig marigold is a good example of a desert plant that is adapted to "harsh" conditions and does not succeed in more "favorable" conditions where it has to compete with other species.

In the Judean Desert, above sea level, there are only small areas where the topsoil is saline. Here *M. nodiflorum* is restricted to the saline patches which are found on chalk and marl on southern slopes, roadsides in which saline soil had been used for paving, and patches rich in nitrogen salts such as corrals of sheep and goats and archaeological sites.

Growth: The five segments of the dry fruits open up when wet by the first winter rains, and the numerous small seeds are washed out by additional rain (see pp. 31-32). Seedling development is slow in winter and accelerates during spring. The size of the plants and the number of flowers and fruits produced depend mostly on the moisture and salt regime at each site. When the plants stop absorbing water from the soil, flowering stops, fruit ripening begins, and the plants turn red. The red color is a result of the loss of chlorophyll which makes other pigments visible.

The property of becoming red when under water stress makes *M. nodiflorum* useful as an indicator of poor soil moisture and high salinity. *M. nodiflorum* flowers between March and June, depending on site temperature. The flowers open only during midday. Most other annuals are not active in late spring and early summer. *M. nodiflorum* uses the water reserves accumulated in the storage tissue during winter when soil water is not as saline as in summer. The plant's use of its water reserves can be demonstrated by cutting plants in flower and leaving them in the open air. Such plants will continue to flower and to produce fruits for at least two weeks.

Other physiologic adaptations of *Mesembryanthemum* species were described previously (see p. 29).

Mesembryanthemum forsskalii Hochst. ex Boiss.
AIZOACEAE
Hebrew: ahal megusham.

Bedouin: samkh.

Name: The species is named for the Finnish botanist P. Forsskal who studied the flora of Egypt more than 200 years ago.

Chorotype: Sudanian, extending into the Saharo-Arabian.

Distribution (Fig. 119): This species is found in warmer parts of the desert than *M. nodiflorum*. It is also a salt tolerant but does not occur in nitrogen-rich sites. *M. forsskalii* stores water in its succulent leaves and stems, and can withstand prolonged periods of drought (Fig. 120).

Growth: Seed dispersal takes place as described above for *M. nodiflorum*. However, the capsules of *M. forsskalii* open faster than those of any other hygrochastic species of Israel and Sinai. Germination and growth take place in winter. Flowers and fruits develop during March and April. Whereas

Fig. 120. A seedling of *Mesembryanthemum forsskalii* with typical succulent leaves.

M. nodiflorum turns red, *M. forsskalii* turns yellow towards the end of its life cycle.

Uses: Bedouin of the 'Arava Valley used seeds of *M. forsskalii* for preparing a black bread. Dry plants collected in summer are put in water and the seeds settle to the bottom of the container. The seeds are washed, dried, and ground into flour. The Bedouin say that this bread is not very tasty but is an important source of food.

Salsola inermis Forssk.
CHENOPODIACEAE
Hebrew: milkheet khumah.

Bedouin: khadhraf.

Name: the species name means "not spiny". The genus name is discussed on page 90.

Chorotype: Saharo-Arabian.

Distribution (Fig. 121): *S. inermis* is a summer annual halophyte. In the Negev Highlands and in the Dead Sea Valley it is a dominant in loess soils where the layer of soil extending from about 20 to 40 cm is saline. Such sites are nearly devoid of winter annuals. It also occurs on stony soils in *Zygophyllum dumosum* or *Artemisia herba-alba* communities. In the summer of 1975, after an exceptionally rainy year, the slopes of the Mitla Pass in western Sinai were covered with

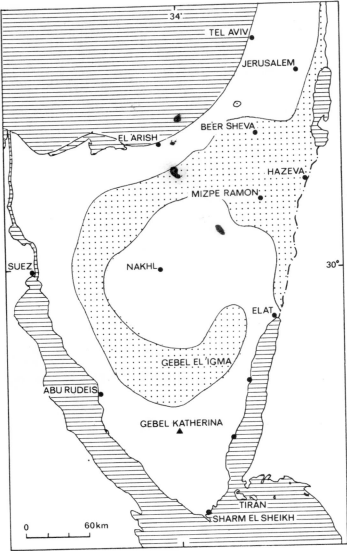

Fig. 121. Distribution map of *Salsola inermis*.

:::::: area where the plant is known to occur but distribution
:::::: pattern is not specified

green stands of *S. inermis*. The abundant moisture presumably leached the salt from the topsoil of these slopes that are otherwise too dry and saline to support dense stands of *S. inermis*.

S. inermis also grows as a colonizer in those disturbed habitats in which soil moisture has not been depleted by winter annuals (94). In parts of the Negev where the top one meter of soil is leached, construction and road building bring deep saline layers to the surface. This provides a suitable habitat for *S. inermis* for a few years until the salts are again leached and non-halophytic annuals take over.

Experiments were carried out to study the response of *S. inermis* (a halophytic summer annual) to growing in proximity to local non-halophytic winter annuals (113). The experiments were done near Sede Boqer under natural rainfall conditions. In the first set of pots, *S. inermis* was sown together with the winter annuals in non-saline soil. The winter annuals successfully completed their life cycle, setting seed

and desiccating toward the end of spring. By that time *S. inermis* desiccated without setting seed. In the second set of pots, *S. inermis* was sown by itself in non-saline soil. Large plants developed but died in summer without flowering. In the third set of pots, *S. inermis* was sown together with the winter annuals in saline soil. Here, the winter annuals failed to develop, while *S. inermis* grew more slowly than in the non-saline soil and had sufficient moisture to complete its life cycle.

Salsola inermis is not generally found in sandy areas, on magmatic or metamorphic substrates, or in gravel plains where the vegetation is restricted to wadis.

Two other local species of *Salsola* have similar life cycles and habitats. *S. jordanicola* is restricted to the northern Dead Sea Valley and the northeastern section of the Judean Desert. The range of *S. volkensii* largely overlaps that of *S. inermis*. Near Kuntilla *S. volkensii* was observed as a dominant in superficial wadis over large areas of chalk.

Growth: Germination takes place after the first winter rains. Up until March, the plants consist of thin cylindrical leaves forming a rosette 1 to 2 cm in diameter (94). In spring the apical bud gives rise to a stem bearing lateral branches and small summer leaves. By this time the winter leaves become dry. Flowers can be found from July to October. Winged diaspores, which look like those of *Hammada* (p. 94) are dispersed by winds in winter. The extent of branching as well as the general size of the plant depend on the available moisture.

Anastatica hierochuntica L.
CRUCIFERAE
English: rose of Jericho.
Hebrew: shoshanat-Yerikho haamiteet.
Bedouin: kmaysha, kaff-e-rakhman, kaff-e-nabbi.

Name: The genus name is derived from the Greek word *anastasis* which means resurrection. The curved dry stems of the dead plant straighten when wet (Fig. 122). For many Christians this symbolizes the resurrection of Jesus. Pilgrims often acquire dry specimens of *A. hierochuntica* as souvenirs

Fig. 122. *Anastatica hierochuntica*: dry dead plant (left) and similar dead plant half an hour after being immersed in water (right). The taproots seen here are not normally exposed in nature.

the Holy Land. Rose of Jericho plants were uncovered by ▪r. Z. Meshel in a chamber of a Jewish temple at Kuntillet ▪jrud in Sinai, built around 3,000 years ago. This suggests ▪at the plant was used in ancient Jewish worship. The Latin ▪ecies name means "from Jericho" and is based on the ▪mmon name in several European languages. The Bedouin ▪ames refer to the fist-shaped dry plant that opens after a ▪bstantial rain. Each rain shower is so greatly valued by the ▪edouin that they call the plant "the hand of the merciful" ▪aff-e-rakhman) or "the prophet's hand" (kaff-e-nabi). The ▪ecies name in Hebrew indicates that this species is the "true" ▪se of Jericho. Another hygrochastic species in Israel, ▪steriscus pygmaeus, has been called "false rose of Jericho".

▪horotype: Saharo-Arabian.

▪istribution (Fig. 123): The rose of Jericho is principally a ▪ant of contracted vegetation. It also occurs in a diffused ▪attern at the base of escarpments along the Dead Sea on ▪ndstone outcrops at Yamin Plain near Dimona, and on ▪mestone slopes near the boundary of the diffused semishrubs ▪getation.

Fig. 124. *Anastatica hierochuntica*, a green plant bearing immature fruits (left) and a dry dead plant from a previous season containing viable seed (right).

Growth: *A. hierochuntica* is a winter annual and is one of the few annuals known to be lignified. Seeds are dislodged by raindrops after the plant is exposed to sufficient rain (see page 31 for details on the water measurement mechanism). Seeds germinate in as short a time as half a day. The size of a mature plant depends on the amount of available water. Some plants are as small as 1 cm in height and bear only two fruits, while the largest plants may reach 50 cm in diameter and bear hundreds of fruits. The flowers are small, off-white in color, and unspectacular. As the fruits ripen, the plant starts to dry. The stems curve inward enclosing the fruits, each of which contains up to four seeds (Fig. 124). The taproot anchors the dead plant to the ground for dozens to hundreds of years (Dr. J. Friedman, pers. comm.). The dead plant does not appear to be subject to decomposition by fungi or bacteria. This may be explained by the arid habitats in which it grows and possibly by some chemical constituents of the stem which inhibit microbial activity.

After a shower wets the dry plant, the curved lignified stems enclosing the fruits become straight within two hours. The cellulose cell walls of the dead tissues facing the center of the plant expand mostly in the direction of the stem axis (43). However, the tissues facing the periphery do not expand in a specific direction. As a result, the stems become straight (Fig. 122). Wetting also softens the sutures of the fruit valves. A second shower may tear some of the sutures, releasing a few seeds. Only a few fruits are opened with each series of showers. When drying, the differential construction of stem tissues causes the stems to curve in again. Thus the seeds from a given plant are released in the course of many years. This mechanism ensures seed dispersal during favorable moisture conditions, while minimizing the exposure of seeds to rodents, ants and other seed-eating animals.

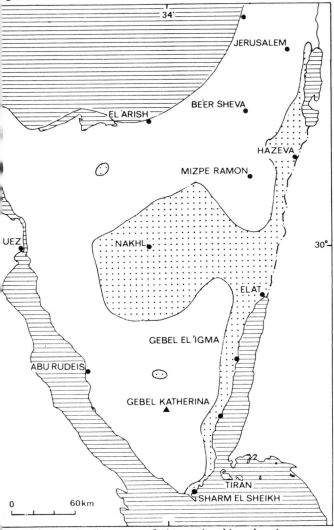

Fig. 123. Distribution map of *Anastatica hierochuntica*.

area where the plant is known to occur but distribution pattern is not specified

Asteriscus pygmaeus (DC.) Coss. et Dur.
COMPOSITAE
English: oxeye, Jericho oxeye.
Hebrew: kokhav nanasee.
Bedouin: gsasah.
Name: The Latin word *aster* and the Hebrew word *kokhav* mean star; the inflorescences have both tube and ray florets and resemble stars. This species is one of the smallest in the genus.
Chorotype: Saharo-Arabian, extending into the Irano-Turanian.
Distribution (Fig. 125): *A. pygmaeus* is mainly a plant of diffused vegetation in limestone and dolomite areas. It is a frequent component of *Zygophyllum dumosum* communities, and in the relatively moist and cool portions of its range is confined to southern slopes. *A. pygmaeus* is infrequent in the sandstone, magmatic, and metamorphic rocks of Sinai.
Growth: *A. pygmaeus* is a winter annual. Dry plants from

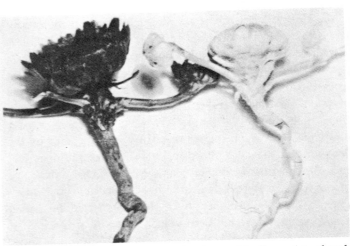

Fig. 126. *Asteriscus pygmaeus.* A plant with three dry closed heads (right); a similar plant five minutes after wetting (left), showing two opened heads. The taproot starts just beneath the branching point.

previous years bear heads consisting of numerous achenes enclosed within stiff involucral bracts. These bracts open within five minutes after wetting, due to the differential elongation of fibers on the inner side of the bracts (Fig. 126). Some of the exposed achenes are released in the vicinity of the mother plant if there is additional rain before the bracts become dry. When dry, the bracts close again until the next rain.

Plant size depends on the available moisture. When the water supply is limited, plants develop only one inflorescence whose size depends on the precise moisture available. When there is abundant moisture for a prolonged period, second-order branches develop beneath the first head (the right plant in Fig. 126). Later on, third-order branches with smaller heads may develop from the second-order branches.

Plants with fifth-order heads are occasionally found. All the heads are appressed to the ground.

SPECIES CONFINED MAINLY TO WADIS

Atriplex halimus L.
CHENOPODIACEAE
English: saltbush, silvery orache.
Hebrew: maluakh kipeakh.
Bedouin: gataf.
Name: The genus name is derived from the common Greek name of another *Atriplex* species. The species name means maritime, referring to its common occurrence along the Mediterranean coast of southern Europe. In Hebrew *maluakh* means having a salty taste and *kipeakh* means tall.
Chorotype: Mediterranean and Saharo-Arabian.
Distribution (Fig. 127): In desert areas *A. halimus* is principally confined to wadis having a silty or chalky substrate. It is especially abundant in wadis of chalk and marl bedrock and is relatively rare on sandy, magmatic, and metamorphic substrates.

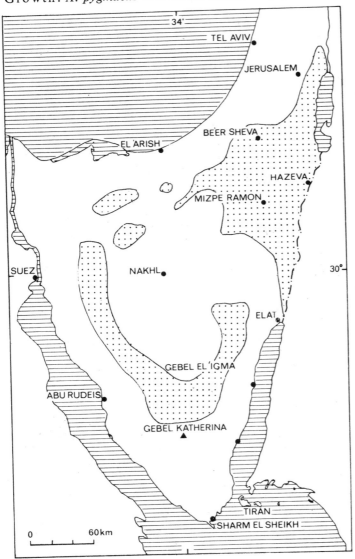

Fig. 125. Distribution map of *Asteriscus pygmaeus.*

⬚ area where the plant is known to occur but distribution pattern is not specified

114

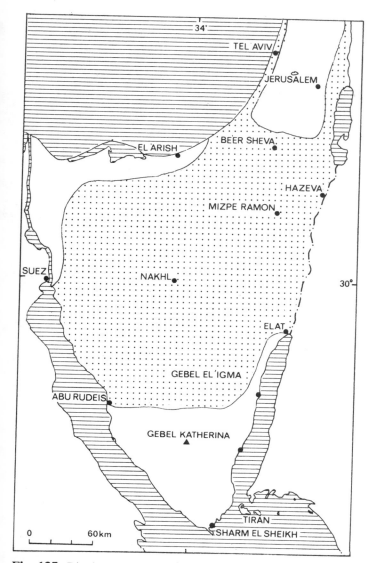

Fig. 127. Distribution map of *Atriplex halimus*.

```
:::::::::  area where the plant is known to occur but distribution
:::::::::  pattern is not specified
```

Growth: *A. halimus* is a shrub up to 2 m high. New stems begin to develop in early spring from buds at the base of older stems. Just as *A. leucoclada*, the plants flower in autumn and the fruits ripen and are dispersed in winter. The number of leaves depends on the available water. Leaves and stems are covered with vesicular hairs. The hairs on young green leaves are filled with saline solution. During summer these vesicular hairs dry out and the leaves become silver-colored, which may function to decrease the absorption of solar radiation.

Ball-shaped galls 2 to 3 cm in diameter frequently develop on young stems. The galls on lignified stems resemble irregular swellings. The leaves of *A. halimus* are an important source of food for the sandrat (*Psammomys obesus*). These rodents, which often burrow holes under saltbush shrubs, have kidneys which enable them to tolerate large quantities of ingested salt.

Uses: The leaves are edible and can be eaten fresh or

cooked in water. The Bedouin use *A. halimus* both for food and forage and have special idioms praising the plant. It has been planted in the Negev for pasture.

Zilla spinosa (L) Prantl
CRUCIFERAE

Hebrew: silon kotsanee.
Bedouin: sillah, silli.

Name: The genus name as well as the Hebrew and Arabic names are derived from the Biblical names mentioned in Ezekiel 2, 6 (*saloneem*) and 28, 24 (*silon*). The species name in Latin and Hebrew refers to its spiny stems.

Chorotype: Saharo-Arabian.

Distribution (Fig. 128): *Z. spinosa* occurs in the relatively warm parts of the Negev and Sinai, mainly in wadis. In southern Sinai it is occasionally found growing in a diffused pattern on slopes of dark magmatic and metamorphic rocks, where the absorption of solar radiation is efficient. Small

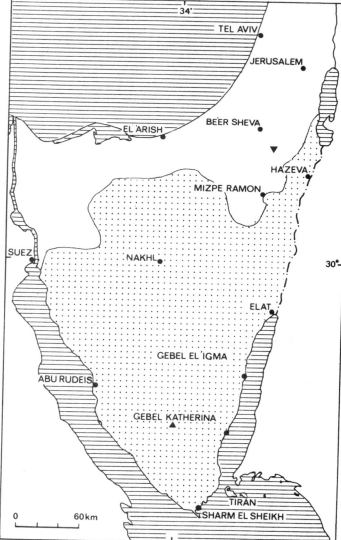

Fig. 128. Distribution map of *Zilla spinosa*.

```
:::::::::  area where the plant is known to occur but distribution
:::::::::  pattern is not specified
```

specimens are found at elevations as high as 2,000 m. It is hard to show on the map the areas of optimal water regime. West of Dimona a few specimens of *Z. spinosa* inhabit a sandy substrate near the road, where there are no competing species and the plant can utilize enough runoff water from the road.

Growth: *Z. spinosa* is a stem-assimilant. The mature shrubs vary from 20 cm to 1.5 m in height and in crown diameter. The leaves of *Z. spinosa* seedlings range approximately from 2 to 8 cm in length and are shed as the plant matures. The leafless green stems carry on photosynthesis (see pp. 29-30). Toward summer, when stem formation and elongation cease, the branches become stiff and their apices become spiny. Sufficient moisture in the rhizosphere promotes the winter growth of new branches bearing small leaves and the development of flowers in spring. The violet or lilac flowers, efficiently pollinated by bees, produce fruits toward summer. The indehiscent fruits remain on the mother plant. Seed dispersal takes place when entire dead plants or broken pieces are dislodged by floods.

In the absence of sufficient water, the plants die two years after germination. Hence, *Z. spinosa* is either a biennial or a perennial, depending on the moisture regime at the particular site.

Uses: This plant is an excellent source of fodder for camels in southern Sinai.

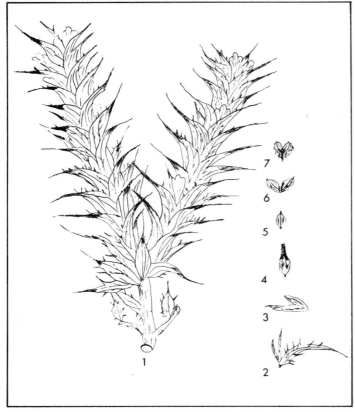

Fig. 129. *Blepharis ciliaris* : 1. flowering branch; 2. spiny bract with two bracteoles; 3. sepals opened after being wetted; 4. fruit enclosed in two dry sepals; 5. fruit; 6. disected fruit with two seeds; 7. germinating seed showing the hairs on seed coat, the opened cotyledons and the emerging root.

Blepharis ciliaris (L.) Burtt
ACANTHACEAE

Hebrew: reesan ne'ekhal.

Bedouin: shok e dhab.

Name: The Latin and Hebrew genus names meaning eyelash and the Latin species name describe the hairs on the seed coat which become straight when moistened. The Arabic name means lizard's tail in reference to the spiny branches. A common synonym is *B. persica* (Bornm.) Kunze.

Chorotype: East Sudanian, extending into the Saharo-Arabian.

Distribution (Fig 129): *B. ciliaris* is found in warmer desert areas, usually in wadis. In southern Sinai it occurs at elevations up to 2,400 m on dark heat-absorbing magmatic and metamorphic rocks. It also occurs outside wadis on sandstone outcrops in Yamin Plain, east of Dimona. In Israel and Sinai there are several populations which differ in their morphology and physiology (59).

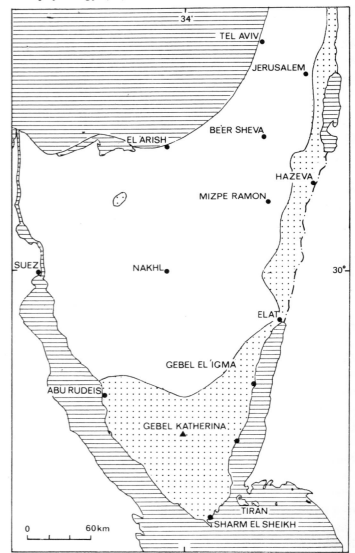

Fig. 130. Distribution map of *Blepharis ciliaris*.

area where the plant is known to occur but distribution pattern is not specified

Growth: *B. ciliaris* is either an annual or a perennial depending on available moisture. The seeds germinate shortly after dispersal (see p. 32 and Fig. 130) and the seedlings possess a pair of crescent-shaped cotyledons. The leafy stems of mature plants become spiny, and flowers are borne in the axils of the spiny bracts.

SUDANIAN TREES

Acacia raddiana Savi
MIMOSACEAE

Acacia tortilis (Forssk.) Hayne

Hebrew: shitah slilaneet.
Bedouin: sayal, siyal.

Hebrew: shitat hasokhekh.
Bedouin: samrah.

Name: The genus name is derived from the Greek *akis* meaning acute, referring to the spiny stipules characteristic of all *Acacia* species except the many Australian species. *A. raddiana* is named in honor of the Italian botanist Raddi; and *tortilis* describes the twisted, fruits of this species (a characteristic shared by *A. raddiana*). The Hebrew name "shitim" is mentioned ten times in the Bible (Exodus 25, 26, and 27) as the only type of wood used for building the Tabernacle by the Israelites in the desert.

Chorotype: Sudanian.

Distribution: (Fig. 131): The distribution and life cycles of the three species of *Acacia* that occur in the deserts of Israel and Sinai were studied extensively by G. Halevy and our description here is based on his studies (61, 62, 63). In this book we refer only to the two most common species. (For details concerning *Acacia gerrardii* subsp. *negevensis*, the third species of the Negev and Sinai, the reader is referred to Halevy's papers).

Although *Acacia* species are characteristic of warmer desert areas, *A. raddiana* occurs in cooler and moister sites than *A. tortilis*. The latter is found mainly in the lower 'Arava Valley and there are occasional individuals on the southeastern coast of Sinai. *A. raddiana* occurs at elevations up to 1,400 m on dark magmatic and metamorphic substrate. The greater resistance of young *A. tortilis* plants to drought was demonstrated by making counts of the *Acacia* trees growing in wadis of different sizes. *A. tortilis* dominates small wadis where conditions are relatively dry, while *A. raddiana* becomes increasingly predominant with increasing wadi size and more favorable moisture regime (62 and Fig. 132). *A. tortilis* trees reach considerable size at 'En Yahav near wells in which the water level is 12 m deep. These trees are in a small wadi (Fig. 30 and p. 48). This indicates that while the moisture available to the young plants will affect the proportion of the two species in the 'Arava, the size of the mature trees is determined by the water regime at deeper soil layers.

In the northwestern Negev, *A. raddiana* trees occur in a diffused pattern in an area where the climate is similar to African savannas, i.e., mean annual rainfall of 150 to 200 mm and mean annual temperature of more than 20°C. However, throughout its range in Israel and Sinai, *A. tortilis* is restricted to wadis.

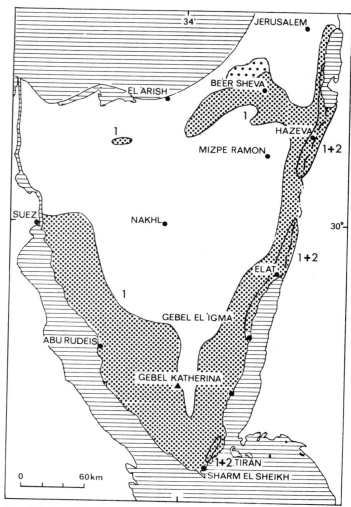

Fig. 131. Distribution map of (1) *Acacia raddiana* and (2) *A. tortilis*.

⋯ area where the plant grows in a diffused pattern

▒ relatively dry area where the plant is restricted to habitats wetter than most of the area (contracted pattern)

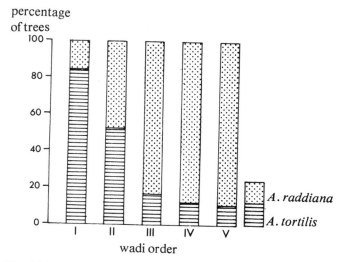

Fig. 132. Relationship between wadi size and the relative proportion of *A. raddiance* and *A. tortilis* trees. The smallest wadis are those of order I and the largest of order V (after 61).

Fig. 133. *Acacia raddiana*, the sole tree growing in a wide wadi in southeastern Sinai.

Seeds are dispersed by many animals that eat the fruit (e.g., gazelle, ibex, rock hyrax, camel, and goat). The partial digestion of the seed coat in the animal intestine increases germinability. Seeds are also dispersed by flood water on which the fruits float.

Growth: The optimal growth conditions of both *A. raddiana* and *A. tortilis* are found in the savannas of East Africa, where rain falls in summer. The life cycle of these *Acacia* trees in Israel and Sinai is geared to the seasonal changes in the East African savanna and not to the local climate. In *A. raddiana* the growth of new stems bearing leaves begins at the end of July. In the African savanna flowering reaches its peak in July, whereas in Israel and Sinai it extends from June to November. In southern Sinai, *A. raddiana* generally flowers around the same date regardless of the annual variations in rainfall. The water reserves in the beds of these wadis are presumably large and the local Bedouin say that *A. raddiana* trees can even survive ten years of drought.

A. raddiana sheds most of its leaves at the beginning of summer. There may also be a total loss of leaves as the result of cold damage in winter, particularly in the northwestern Negev.

A tortilis, which requires higher temperatures throughout its life cycle, has shorter flowering and growth periods than *A. raddiana*. The optimal temperature for germination of *A. tortilis* is between 25° and 35°C, while for *A. raddiana* it is between 20° and 30°C (61).

Uses: Leaves, inflorescences, and fruits of *Acacia* are important sources of forage for sheep and goats, and for this reason the Bedouin of Sinai do not cut these trees for fuel. Because the production of *Acacia* forage is relatively independent of annual moisture conditions, it constitutes an extremely reliable and valuable resource. Trees of *A. raddiana* injured by insects secrete a gum used by the Bedouin in Sinai as a base for candy. Fibers from the bark of roots and trunks can be used for preparing rope.

Ziziphus spina-christi (L.) Desf.
RHAMNACEAE
English: Syrian Christ-thorn.
Hebrew: sheyzaf matsuy
Bedouin: sidir.

Name: The genus name in Latin is probably derived from the Hebrew name. The species name refers to the legend that the thorns of this tree were used by the Romans for the crown in the Crucifixion. This species name was also applied to *Paliurus spina-christi* which grows in the Galilee. There are churches in Israel where the statue of Christ is decorated with a crown of *Sarcopoterium spinosum*, a very common semishrub around Jerusalem. Since the Romans wished to humiliate Jesus, it seems more likely that they used the unimportant semishrub

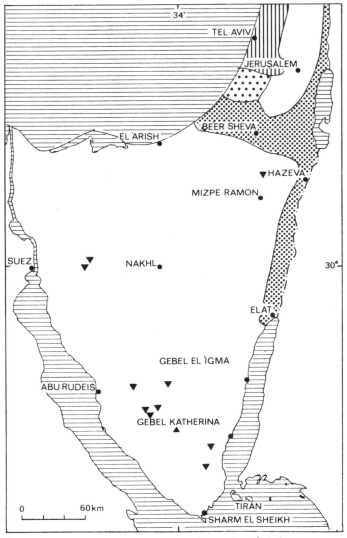

Fig. 134. Distribution map of *Ziziphus spina-christi.*

area where the plant grows in a diffused pattern

relatively dry area where the plant is restricted to habi-tats wetter than most of the area (contracted pattern)

relatively wet area where the plant is restricted to habitats drier than most of the area

▼ one or a few individuals

growing nearby rather than bringing a plant from Jericho.
Chorotype: Sudanian, extending into the Saharo-Arabian, the Mediterranean, and tropical Africa.
Distribution (Fig. 134): Like *Acacia, Z. spina-christi* is a plant of African savannas, but is more cold-resistant and requires more water. In Israel, the Syrian Christ-thorn is found further north than *Acacia*. However, their ranges overlap in the northeastern Negev, where *A. raddiana* occurs in a diffused pattern and *Z. spina-christi* is restricted to wadis. *Z. spina-christi* grows in a diffused pattern on the Mediterranean coastal plain between Tel-Aviv and Ashqelon, and near the Sea of Galilee on the basaltic slopes of the Golan and Lower Galilee. In moister areas it grows in places where *Quercus ithaburensis* and *Ceratonia siliqua* were removed for fuel or to prepare the land for ·cultivation. Once established, *Ziziphus* trees are very resistant, even to fire. Fire resistance is probably an adaptation to savanna habitats which are subject to frequent fires. In the Jordan Valley, south of the Sea of Galilee, it mostly grows in wadis and near fresh water springs. In the oases of Jericho, 'Auja, and Jiftlik in the Lower Jordan Valley, it is accompanied by *Balanites aegyptiaca*. *Z. spina-christi* trees are common in wadis in the northern 'Arava Valley, with dense populations at Nahal Shezaf near 'En Yahav. The largest and most famous tree in the 'Arava Valley is a *Z. spina-christi* specimen growing at a fault line spring near 'En Hazeva. When the water table of the Hazeva area was lowered following intensive agricultural development, water was artificially supplied to rescue this old tree. *Z. spina-christi* is planted as a fruit tree by Bedouin in southern Sinai, who also look after wild specimens growing in wadis. A few wild trees occur in northwestern Sinai near the Mitla Pass.
Growth: *Z. spina-christi* blooms several times a year in the warmer parts of its range, e.g., the Jordan Valley and the Dead Sea Valley. Pollination was studied by Galil and Zeroni (46). In cool years and in the cooler parts of its range in Israel and Sinai, leaves fall in the winter and new leaves develop in spring. Flowering and fruit development take place mainly in summer. The fruit has a juicy sweet mesocarp, and is eaten by various animals which excrete the hard-coated seeds.
Uses: The fruits, called "dom" or "nabq" in Arabic, are eaten fresh or dried. The Bedouin of southern Sinai preserve the fruit by drying and prepare cakes from the ground dry mesocarp.

Phoenix dactylifera L.
PALMACEAE
English: date palm.
Hebrew: tamar matsuy.
Bedouin: nakhlih.
Name: The scientific genus name is the Greek common name; *dactylifera* means furnished with fingers. The Hebrew name is mentioned many times in the Bible.
Chorotype: Sudanian, extending into the Saharo-Arabian.
Distribution (Fig. 135): Date palms are cultivated in warm countries throughout the world. In Israel and Sinai wild

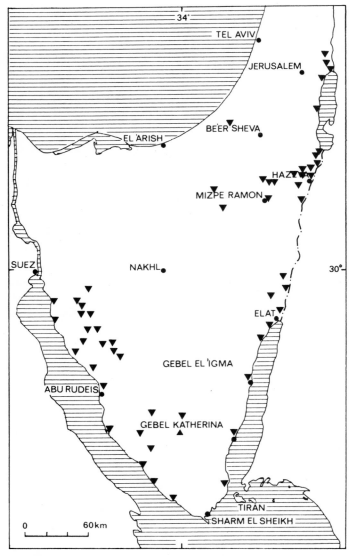

Fig. 135. Distribution map of *Phoenix dactylifera*. Only those trees which are presumed to have germinated and became established spontaneously are included.

▼   one or a few individuals

populations occur at desert springs, which are considered the natural habitat of the species and the source of stock first used for domestication some 6,000 years ago (111, 142, 148). Wild date palms have many trunks (Fig. 50 and Plate 12) and are usually accompanied by *Juncus arabicus*. They bear small tasty fruits with relatively large seeds that are dispersed by birds and humans.

Seeds germinate on wet soil near desert springs and roads, where the soil may remain at temperatures above 20°C for at least two to three months. Only those palms growing near springs will reach maturity. Frequently, wild date palms can be found growing on wet saline soil with a salt crust. However, the roots penetrate to less saline layers beneath. Presumably, germination took place when conditions were less saline.

The *Washingtonia* palm in southern California, like the date palm in the Middle East, also occupies desert springs.

Cultivated palms usually have single trunk because the lateral buds are deliberately cut off. To avoid high salt stress near the soil surface, they are planted in deep holes where their roots are exposed to the less saline water table.

Date palms in desert oases have played a significant role in Jewish history and culture. This is reflected in the many Biblical passages that refer to the date palm. The Israelites depended heavily on date palms found in oases during their wandering in the desert after fleeing Egypt. The date palm is used as a symbol in religious ceremonies commemorating the Exodus. The famous Roman coin depicting the destruction of the Jewish Kingdom in Israel ("Judaea Capta") shows a humiliated woman, representing the Jewish people, under a date palm tree.

Growth: The date palm has a trunk that only produces branches near the base. When these branches are not removed, they grow to have the same diameter as the original trunk. The pinnately compound leaves are 2 to 3 m long at maturity. The trunk at the base of each leaf is covered with fibers.

The date palm is dioecious, i.e., male and female flowers develop on separate trees. Flowering takes place in spring, pollination is by wind, and the fruits ripen in autumn. An old Talmud story tells of a barren female tree in Khamatan that "longed" for a male palm in Jericho, and produced fruits only when pollinated by flowers brought from that male tree (Midrash Raba, Bamidbar 3, 1).

Uses: This is the most useful tree in the desert. It indicates sources of fresh water; is has edible fruit; the trunk and dry leaves are used for building houses, huts, and fences for animal enclosures; fibers on the trunk and in the leaflets are used for making rope (Figs. 143 to 148); leaflets torn into 5 to 10 cm wide strips which are made into baskets; the peduncles of the branched inflorescence are woven into sieves; and hollowed-out trunks are cut lengthwise and used as permanent water conduits for irrigation.

CHAPTER 5. USEFUL PLANTS

People living in the desert **have always** been dependent on plants for their survival. Plants may indicate the presence of drinking water or may themselves be a source of water. They are a source of food for man and his livestock, and supply the raw materials for making mattresses, shelters, ropes, and many other things. It is also essential to be able to recognize which plants are poisonous.

INDICATORS OF POTABLE WATER

The availability of water is critical for human survival in the desert. Some plants which prefer or require a high water table can be detected from a distance and therefore used to locate water near the surface (Table 3). The water may be fresh, slightly saline or too saline to be potable. Species useful for detecting potable water include those which require fresh water

Plate 16.　An opened ancient cistern in the Negev Highlands filled with runoff water accumulated from the slopes. Seeing the *Tamarix nilotica* tree from a long distance, a viewer could guess that a cistern is located on the slope.

Table 3. Plant indicators for water quality detected from various distances

water quality	Plants recognizable from up to 100 m	Plants recognizable from up to several kilometers (with the aid of field glasses)
1. fresh water	*Typha australis, Cyperus distachyus, Holoschoenus vulgaris, Mentha* spp.	*Saccharum ravennae,* green algae, *Salix* spp.
2. fresh to slightly saline water	*Polypogon monspeliensis*	*Phoenix dactylifera, Populus euphratica, Phragmites australis, Arundo donax*
3. fresh to saline water	*Juncus arabicus, Imperata cylindrica, Alhagi maurorum*	*Tamarix* spp., *Nitraria retusa*
4. saline water	*Tetradyclis tenella, Zygophyllum album, Arthrocnemum macrostachyum, Zygophllum simplex, Sphenopus divaricatus*	*Suaeda monoica, Halocnemum strobilaceum*

Fig. 136. Inflorescences of *Phragmites australis*.

121

Fig. 137. *Juncus arabicus* growing on muddy soil near a spring in the desert.

and those which can tolerate slightly saline water (groups 1 and 2 in Table 3). Some plants need relatively fresh water for germination and establishment but can tolerate much more saline conditions as mature plants (group 3 in Table 3). Therefore, they indicate that the water in that area was once fresh but may now be saline. Hydrohalophytes require some salinity for germination and can tolerate much higher salinity. These halophytes will therefore serve as indicators for unpotable water (group 4 in Table 3).

FINDING SPRINGS

On the way from Sede Boqer to 'En Mor one passes two rows of *Tamarix nilotica* trees, one row on each side of Nahal Zin. These trees indicate the presence of a high water table in the area. Further south are *Nitraria retusa* shrubs, tufts of *Juncus arabicus* (Fig. 137) and occasional clumps of *Phragmites australis* (Fig. 136). These species indicate moist conditions but there may not be drinking water throughout the year. Further upstream, near the parking lot of the canyon nature reserve, is a dense clump of *Phragmites australis* along the wadi channel and a *Populus euphratica* tree. The *Phragmites* reach a height of 3 to 4 m at the center of this clump, but are smaller at the margins where they are accompanied by *Juncus arabicus*. The tall *Ph. australis* indicates the likelihood of finding potable water here. Near the margins of the clump, the water is likely to be brackish. Near

this site can be found remnants of plants carried here by floods from upstream. Among these plant remnants are leaves of the hydrophyte *Typha australis*. These leaves are about 1 cm wide and in cross section reveal a system of minute air chambers. The presence of *Typha* leaves here indicates the existence of fresh water further upstream (Plate 5).

The 'En 'Avdat springs also support the growth of *Cyperus distachyus, Polypogon monspeliensis, Populus euphratica,* and *Apium graveolens* near the open water. Green algae develop in open non-stagnant fresh water. Many of these algae become white upon drying due to the deposition of lime. The white crust of algae indicates that water flowed and passed here during the rainy season and there is probably a source of water further upstream.

A combination of two plant indicators can be used to locate water in the extremely dry El Q'a area near Wadi Jiba, 38 km north of Et Tur. The vegetation of the alluvial plain is very sparse. However, remains of *Phragmites australis* or *Arundo donax* stems are found on plants in the wadis. These are fragments of hydrophytes growing near springs or in sites with a high water table further upstream in the canyons of the magmatic mountains. Remains of algae on the gravel at the mouth of Wadi Jiba provide even stronger evidence of potable water. A few meters upstream there is indeed a small natural pool with fresh water. Similar water pools and plant indicators occur in several other wadis descending from the southern Sinai mountains.

A salt marsh with *Halocnemum strobilaceum* west of El 'Arish

In an area of the Dead Sea Valley 1 to 5 km southwest of the potash factory called the "Dead Sea Works", underground water approaches the soil surface at many points along the geological faults. In places where rainfall is the only source of water, the Lisan Marl here supports vegetation only in wadis. Plants growing outside of the wadis are dependent on another source of water. For example, large plants of *Prosopis farcta*, *Nitraria retusa* and *Tamarix* grow on hills of Lisan Marl where they are supplied by water from springs along the faults. Sites underlain by substantial water have relatively dense vegetation dominated by *Juncus arabicus* and *Phragmites australis*. A few spontaneous date palms are also found here. Except in occasional wet years, there is no direct outflow of water at these sites. The patches of dense vegetation on the Lisan badlands grow in strips along the faults.

The same plants which help us locate springs in the Negev can be used for this purpose in northern Sinai. In southern Sinai, where magmatic and metamorphic rocks predominate, the most common indicator plant is *Holoschoenus vulgaris*. *Juncus punctorius* and *Mentha longifolia* also occur in sites where fresh water is available for much of the year. The white strips associated with algal growth are particularly prominent on the dark gravel and can be observed from a distance of a few kilometers.

LOCATING WELLS

Shallow wells that can be dug by hand ("thamila" in Arabic) are easily located by plants mentioned as indicators for springs. However, there are seasonal fluctuations in water depth, and water may be absent in dry years.

Deep wells dug by inhabitants throughout the ages exist at various points in the deserts of Sinai and Israel. Large *Tamarix aphylla* trees are often found near deep wells in northern Sinai and the Negev.

CISTERNS

Cisterns for collecting runoff water have been dug by residents of the Negev for thousands of years and continue to function if the feeding channels are intact. *Tamarix nilotica* trees that develop from wind-borne seeds can sometimes be found growing in abandoned cisterns (Plate 6). They become established in the mud accumulated at the bottom of the cistern. The presence of these trees on slopes that otherwise support only semishrubs almost always indicates the existence of a cistern. Ancient cisterns can also be located from a distance by the light-colored rock debris which was piled on the slope when the ancient inhabitants dug the cistern.

SALT MARSHES AND COASTAL AREAS

Water that infiltrates sand and gravel flows underground toward the lowest elevations in the area. Where the water is near the surface but sufficiently deep to escape salinization through evaporation, fresh water may be found. At such places near the coast of the Gulf of Elat, the Bedouin have dug wells at Taba, Nuweiba, Dahab and Nabq.

Zones of halophytic vegetation often occur in depressions moistened by underground water (Plate 7). Succulent Chenopodiaceae, such as *Arthrocnemum*, *Halocnemum*, or *Suaeda* spp. dominate most of these saline zones (group 4 in Table 3).

DISTILLATION OF WATER FROM PLANTS

Water stored in the tissues of many plants may not be available to humans because of the salts and other unpalatable substances they contain. Potable water can be obtained from these plants by means of distillation as follows. A pit is dug in the soil, one meter in diameter and half a meter deep. A clean container is placed in the center of the pit. The pit is then half-filled with green plants distributed around the container. (Succulent plants have a thick epidermis and therefore need to be crushed before using). The pit is then covered with a transparent 1.5 m x 1.5 m polyethylene sheet held down by soil at the margins. A stone is placed at the center of the sheet forming an inverse cone, the tip of which should be about 5 cm above the opening of the container. The heat of the sun penetrates the polyethylene, thus warming the air and plants inside ("the greenhouse effect"). Heat causes the evaporation of water from the plant tissues, and the resulting water vapor condenses on the inside of the polyethylene sheet. The condensed drops of water descend along the cone and drip into the container. A single pit can yield 1.5 liters of water per day. The system should be kept covered for at least five hours of bright daylight in order to accumulate a reasonable amount of water. This technique can be adopted to distill drinking water from salt marsh mud, sea water, and water from saline springs.

Not all plants are equally suited for distillation. Poisonous plants may not be used because of the danger that volatile substances may condense with the water. Also, tamarisk stems should not be used because they tend to become muddy and pollute the water.

EDIBLE PLANTS

Leaves of most palatable desert plants should be cooked, but some can be eaten raw. Leaves, young stems, and inflorescences of *Scorzonera papposa* can be eaten as a raw salad. Young leaves of *Atriplex halimus, Malva parviflora, M. nicaeensis,* and *M. sylvestris,* seedlings of *Eruca sativa, Erucaria boveana,* and *Reboudia pinnata,* leaves and young fruits of *Rumex vesicarius* and *Rumex cyprius* can all be consumed raw in small amounts. However, *Rumex* leaves contain oxalic acid and damage the kidneys if eaten in quantity. The base and central vein of young *Gundelia tournefortii* leaves can be eaten raw or cooked. Young stems and central veins of young leaves of *Notobasis syriaca* and *Silybum marianum* can be eaten raw as a source of food and moisture.

Some plants are more palatable after boiling, and the juice may be discarded, if necessary, to improve the taste. Cooking the leaves is recommended for the following species: *Atriplex* spp., *Malva* spp., *Centaurea* spp., *Cichorium pumilum, Urtica* spp., *Moricandia nitens, Diplotaxis harra, Sisymbrium irio,* **Eryngium creticum,** and **Rumex** spp. The immature inflorescences of *Gundelia tournefortii* have a taste when boiled similar to artichoke.

Roots of *Scorzonera papposa* (Fig. 138), *Emex spinosa,* and

Fig. 138. *Scorzonera papposa.* The roots, leaves and inflorescences may be eaten raw or roasted.

Eryngium creticum, and young root-tubers of *Erodium hirtum* can be eaten without cooking. Bulbs of *Tulipa amblyophylla,* a preferred food of porcupines, are palatable for humans. However, this plant is a protected species and should not be collected for food.

Thick roots of *Scorzonera papposa* and tubers of *S. judaica* can be roasted in a bed of coals.

Truffles, the tuber-like fruiting bodies of certain subterranean ascomycetous fungi, can be found in spring in the sandy and loess soils of the Negev. They are usually associated with the small semishrubs *Helianthemum sessiliflorum* in sandy soils, and with the annual *H. ledifolium* in sandy and loess soils (see p. 19). Truffles can be roasted in a bed of coals, or cut into pieces and fried in oil with onions. Bulbs of *Allium ampeloprasum* can be substituted for onion. Truffles are collected and sold commercially in Israel under their North African name "tarfaz". the Bedouin name is "kama" or "kmama", which is similar to the ancient Hebrew name "kmeheen".

FOOD SEASONINGS

Salt can be obtained from the secretions of stem or leaf glands from several plants. *Reaumuria negevensis, R. hirtella, Tamarix* spp., *Avicennia marina, Frankenia* spp., and *Cressa cretica,* all, secrete solutions which crystallize to salt when dry. Soaking the leafy stems in water results in a saline solution that can be used as is or evaporated to produce crystallized salt. Leaves of *Atriplex halimus* have a high salt content during summer and can be added directly to food for seasoning.

Apart from dates, there is no wholly satisfactory source of sugar in desert plants. The sugary fluid "manna" from *Hammada salicornica* and *Anabasis setifera* is only available in summer, and in most years, only in small quantities (see p. 95, Fig. 116 and Plate 15).

Young stems of *Tamarix nilotica* may be covered in summer with a sweet fluid excreted by scale insects (Fig. 139). This insect sucks the plant sap and excretes a sugary fluid which is hard to collect. Sugar crystals or drops containing

Fig. 139. *Tamarix nilotica.* The droplets of honey dew in the axils of the lateral twigs are excretions of scale insects.

sugar can also be found on flowering buds and young leaves of several *Capparis* species during summer.

Important herbs used in both oriental and occidental cooking are thyme, marjoram, and oregano. The characteristic flavors in these herbs are largely due to the presence of the phenols thymol and carvacrol in the essential oils of these plants. The leaves can be used fresh, dried and crushed, or cooked in oil. Flavored margarine can be made by adding the leaves to melted margarine, heating the mixture for a short while, and then cooling the mixture to make it hard again. The species in the deserts of Israel and Sinai that contain thymol and carvacrol and can be used for flavoring food include the following Labiatae: *Majorana syriaca* (Districts 1, 4, 11 and 19); *Corydothymus capitatus* (Districts 1 and 5); *Thymus boveanus* (Districts 3, 4, 11, and 12); *Thymus decussatus* (District 19); *Origanum isthmicum* (District 11) and *Satureja thymbrifolia* (District 1).

Young leaves and inflorescences of *Foeniculum vulgare* (found in Districts 3, 4 and 19) can be used to flavor salads and cooked dishes. Another Umbelliferae, *Pituranthos tortuosus* (Districts 3, 4, 5, 6, 9, 11, 12 and 19) has a taste similar to that of celery and can be used to flavor soup.

The flowering buds and young fruits of *Capparis aegyptiaca* and *C. cartilaginea* can be pickled to produce the famous delicacy "capers". The flowering buds, young fruits, and young leaves of these species are put into water saturated with salt. The liquid is replaced after a week and then again after a second week. Once the bitter taste of the plant is removed, the capers are put into a solution of 2/3 vinegar and 1/3 water with salt added to taste. The capers are ready after one or two weeks in the vinegar solution. Their taste improves with time. *C. cartilaginea* capers are sharper in taste than *C. aegyptiaca*, while capers of *C. spinosa*, a Mediterranean species, have the mildest flavor. Leaves and stems of *Foeniculum vulgare* and *Ridolfia segetum* can be added to the vinegar solution for additional flavor.

Several scented desert plants can be used to make a drink or to flavor tea by steeping the appropriate parts in boiling water. Each of these plants has a characteristic taste. They may be bitter in excess, so that the plant parts should be added gradually to determine the proper amount to use. "Tea" made from many of these same plants are used by Bedouin to relieve stomach pains, headaches, and coughs. Plants whose leaves and stems are used for these various purposes include: *Mentha longifolia* (Districts 18 and 19); *Origanum dayi* (District 3), *Origanum ramonense* (District 4), *Teucrium polium* (the entire region except for Districts 8, 10, 14, 15 and 16); *Teucrium leucocladum* (districts 15, 17, 18 and 19); *Teucrium pilosum* (Districts 18 and 19); *Salvia multicaulis* and *Ziziphora tenuior* (District 19); *Pulicaria desertorum* (Fig. 118); *Artemisia herba-alba* (Fig. 62); *Cymbopogon parkerii* (District 15); *Cotula cinerea* (District 8); and *Matricaria aurea* (Districts 1 to 7, 11, 12, 18 and 19). The roots of *Rheum palaestinum* (Fig. 140, District 4) have been used for centuries as diarrhetic medicine and "tea", but tea made from the roots of this plant should not be drunk in large quantities. The roots of *Asphodelus microcarpus* (Districts 1, 3, 4, 5, 6, 8, and 11) can be used to make "tea" when the roots are yellow to orange inside and not brown. The Bedouin of southern Sinai make "tea" from the lignified branches of *Rhus tripartita* (Fig. 113).

Fig. 140. *Rheum palaestinum.* The roots may be used to prepare a tea-like beverage and the young inflorescences can be eaten raw.

JUICY PLANTS

Most of the water-storing desert plants of Israel and Sinai are distasteful and not suitable for extracting drinking water. However, the roots of *Emex spinosa* and *Erodium hirtum* (see above) are more important as sources of moisture than as food. Young inflorescences of *Rheum palaestinum* (related to the cultivated rhubarb) are juicy and have a sweet-sour taste. Like the related genus *Rumex*, *Rheum* may be high in oxalates and should not be eaten in large quantities. The thistles *Notobasis syriaca*, *Silybum marianum*, and *Scolymus maculatus* are found mainly at the margins of the desert, and their stems are a good source of liquid.

FRUITS AND GALLS

Fresh fruits of the following species can be eaten: *Ziziphus spina-christi*, *Lycium shawii*, *Ficus pseudosycomorus*, *Nitraria retusa*, *Crataegus aronia*, *Crataegus sinaicus*, date palms, and the sweet pulp surrounding the seeds of *Capparis aegyptiaca, C. cartilaginea,* and *C. decidua.* The green fruits of *Pistacia atlantica* and *P. khinjuk* contain small palatable seeds. *P. atlantica* fruits are sold in the markets of Jerusalem and are used to flavor cakes. Fruits of the following herbaceous plants can be eaten before they become dry: *Hordeum spontaneum, Pisum fulvum, Gundelia tournefortii, Malva parviflora,* and *M. sylvestris.*

Galls on young branches of *Salvia dominica* (Districts 1 to 5) have a pleasant scent and sweet taste when the plants are flowering. The Bedouin call this plant "Khokh" meaning peach, in reference to these galls.

POISONOUS PLANTS

The desert traveler should be able to recognize those inedible and poisonous plants that resemble useful plants. The leaves with undulate margins of *Urginea undulata* resemble the undulate leaves of *Tulipa amblyophylla.* However, the dry leaf

Fig. 141. *Hyoscyamus albus.* A poisonous plant growing in wadis in extreme deserts.

bases subtending the bulb in *Tulipa* are brown and are hairy on their adaxial surfaces; those of *Urginea* are glabrous and light-colored. Also, *Tulipa* has one to three leaves, whereas *Urginea undulata* has at least five leaves per bulb.

The leaves and corms of *Colchicum ritchii* and *C. tunicatum* somewhat resemble the leaves and bulbs of *Tulipa* species. However, the *Colchicum* leaves are not undulate and the adaxial surfaces of the brown dry leaf bases subtending the corm are glabrous. The corm of *Colchicum* species may be poisonous to humans although porcupines eat them in summer. Colchicine, a poisonous alkaloid extracted from the corms and seeds of *Colchicum* spp., is used in genetic research to induce chromosome doubling and medicinally to treat gout (12).

The leaves of *Ferula* species resemble those of *Foeniculum,* but the latter (a palatable plant) has a pleasant scent whereas *Ferula* has an unpleasant odor and should not be eaten. Various medicines are made in Afghanistan, Turkestan, Iran, and Tibet from the roots of *Ferula* species (124).

Many desert plants belonging to the families Solanaceae, Boraginaceae, Asclepiadaceae and Apocynaneae are poisonous, and it is best to avoid all desert members of these families. Most of these plants are not grazed by animals. Among the Solanaceae, *Hyoscyamus* species have an attractive flower (Plate 13) and a distinctive calyx (Figs. 141 and 142). They contain the extremely poisonous alkaloids hyoscyamine, hyoscine, and atropine, used in very low quantities in eye and nerve medicine. Parts of this plant, when in contact with the eye, can cause the pupil to dilate. Impaired vision may last two to three weeks. Several cases of atropine poisoning in Sinai have been recorded, including near-fatal incidents in which the plant was used by tourists to induce euphoria.

Peganum harmala (Zygophyllaceae) is a source of several substances used in medicine including harmine, a sedative prescribed to improve the mood of patients. In the wild, this plant should be regarded as poisonous.

The milky sap of *Calotropis procera* caused an acute inflammatory reaction when splashed into a child's eye (8). The consumption of *Rumex cyprius* and *R. vesicarius* leaves in large amounts may cause urinary bleeding.

MEDICINAL PLANTS

As discussed above, useful medicines are derived from many plants that are generally considered poisonous. This section presents some guidance on the field preparation of medicines that may be useful as first aid remedies.

Dust and other particles in the eye can often be removed by placing a young stem of *Andrachne aspera* (Districts 2, 7, 14, 15, 18 and 19) under the eyelash. This treatment results in a flow of tears that may help to flush out any foreign particles.

The tea-like drinks discussed earlier in this chapter may be useful for relieving stomachaches. Some of the plants are sold for medicinal purposes in the Bedouin market in Be'er Sheva.

Fig. 142. *Hyoscyamus reticulatus.* A poisonous weed in plowed loess soil.

Fagonia mollis, a widespread semishrub, is useful for treating superficial wounds in which the skin has been removed. Green leaves are burnt to produce ash which is applied to the open wound, resulting in rapid healing.

PLANTS FOR MAKING MATTRESSES AND SHELTERS

Comfortable mattresses can be made from several desert plants that are neither spiny nor scented and are low in water content. Suitable species are: *Polygonum equisetiforme, Panicum turgidum, Stipagrostis scoparia, S. plumosa, Pennisetum divisum, Hyparrhenia hirta, Retama raetam,* and *Pituranthos triradiatus.* To prepare the mattress, lay thin non-woody stems parallel to each other so that they cover the ground. Then put a second layer of stems crosswise on top of the first layer. Two such layers may be sufficient, but further layers may make for greater comfort.

Plants have long been used to make huts and other desert shelters. The Bible recounts that the Israelites used dry date palm leaves for huts during the Exodus, and Bedouin use similar desert shelters even today. The Bedouin between El 'Arish and Gaza make huts from the stems of *Artemisia monosperma.*

IGNITING A FIRE WITH DESERT PLANTS

Fire is indispensable to survival in the desert, and the traveler may be interested in knowing how to start a fire without matches. During daylight hours a magnifying glass or field binoculars can be used to concentrate the sun's rays in order to ignite an inflammable material. The smoldering material can then be used to ignite thin stems and branches. A fire can also be started by striking flint with steel to produce sparks, a technique not easily mastered. The bark of *Phagnalon* spp., stem galls of *Artemisia herba-alba,* and inflorescences of *Aerva persica* are all useful sources of wool-like plant hairs that are easily ignited by concentrated sun rays or by sparks. The "wool" can be obtained from the stems of *Phagnalon rupestre* (Districts 1 to 7 and 11 to 13), *Ph. barbeyanum* (Districts 7, 9, 15, 18 and 19) and *Ph. nitidum* (Districts 18 and 19). *Ph. sinaicum* (Districts 18 and 19), has no wool on its bark. Galls of *Artemisia herba-alba* and old inflorescences of *Aerva persica* burn faster than the bark hairs of *Phagnalon.* The ignited "wool" is placed inside a cluster of stems of dry annuals or a cluster of thin woody stems. By blowing the smoldering wool surrounded by thin stems, an expert can start a fire in less than five minutes.

ROPES

Desert inhabitants make ropes from plants by a method used for many thousands of years. Ropes made from date palm fibers by the Bedouin of Sinai are very similar to ropes uncovered in the pyramids of Egypt and in many archaeological sites in the deserts of Israel.

The method is shown in Figs. 143 to 148. First take 5 long date plam leaflets and split them into fine fibers (Fig. 143); twist the fibers and keep twisting them always in the same direction (Fig. 144). After sufficient twisting, the bundle of fibers will bend and the two parts will start winding around each other (Fig. 145). The point of bending should be held firm either by another person or by stepping on it (Fig. 146). Continue twisting the fibers in the same direction at the same time that the two bunches of fibers are wound around each other. Maintain an angle of about 60° between the two parts of the rope. Add more fibers, derived from two leaflets, as soon as the two parts become shorter than 15 cm. The additional fibers are folded in a "V" shape, and each arm is added to one part of the rope (Fig. 147). The two parts are tied together when the rope is the desired length (Fig. 148).

A quicker method, used in many parts of the world, is shown in Figs. 149 and 150. In this way an expert can prepare one meter of rope within two minutes, but it is hard to explain in words.

Using either method, ropes and strings of any diameter may be prepared; sewing thread, shoe-laces, as well as cables for towing a vehicle. The strength of the rope depends on the material used, the expertise of the maker, and the diameter of the rope. Plant materials used for rope should have strong fibers that can easily be separated and fold without breaking. Suitable plant material for ropes includes the following: leaves of *Typha* spp., *Juncus* spp., and *Imperata cylindrica;* stems and leaves of *Scripus* spp. and *Cyperus* spp.; stems of *Avena sterilis* and *Chrysanthemum coronarium;* bark fibers of *Thymelaea hirsuta, Acacia raddiana, A. tortilis,* and *Colutea istria.*

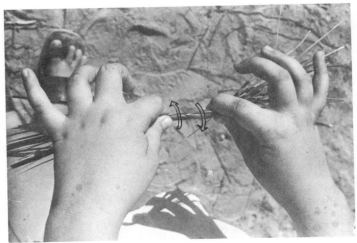

Fig. 144. Twisting the fibers.

Fig. 143. Splitting a date palm leaflet into fine fibers.

Fig. 145. The bundle of fibers bends and the two parts start winding around each other.

Fig. 146. Further twisting of the fibers while the rope is held firm.

Fig. 147. Adding new fibers to lengthen the rope.

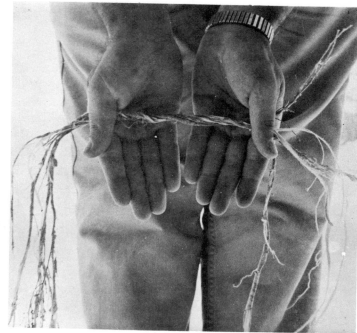

Fig. 149. Bark fibers of *Thymelaea hirsuta* ready for twisting into rope.

Fig. 148. The completed rope.

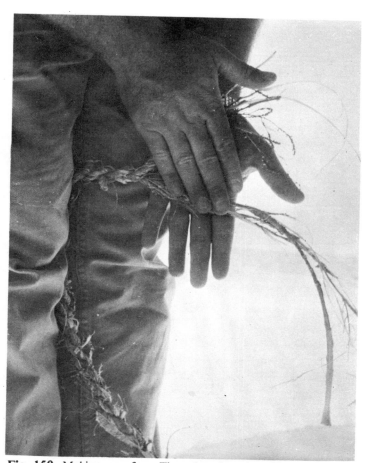

Fig. 150. Making rope from Thymeluea *hymelaea hirsuta* fibers.

REFERENCES

1. Aaronsohn, A. 1910. Agricultural and Botanical Exploration in Palestine. USDA, Washington Bull. 180.
2. Aharonson, Z., J. Shani & F.G. Sulman. 1969. Hypoglycaemic effect of the salt bush *(Atriplex halimus)* — a feeding source of the sand rat *(Psammomys obesus)*. Diabetologia 5: 379-383.
3. Atlas of Israel. 1970. 2nd ed. Department of Surveys, Ministry of Labour, Jerusalem.
4. Bagnold, R.A. 1941. The Physics of Blown Sand and Desert Dunes. London.
5. Bailey, C. & A. Danin. 1980. Bedouin plant utilization in Sinai and the Negev. Economic Botany.
6. Bartov, J. 1974. A Structural and Paleogeographical Study of the Central Sinai Faults and Domes. Ph. D. thesis. The Hebrew University, Jerusalem (in Hebrew with an English summary).
7. Begin, Z.B., A. Ehrlich & Y. Nathan. 1974. Lake Lisan and the Pleistocene. Precursor of the Dead Sea. Survey of Israel Bull. 63: 1-30.
8. Biedner, B., L. Rothkoff & A. Witztum. 1977. *Calotropis procera* (Sodom apple) latex keratoconjunctivitis. Israel J. Med. Sci. 13: 914-916.
9. Black, C.C. 1973. Photosynthetic carbon fixation in relation to net CO_2 uptake. Ann. Rev. Plant Physiol. 24: 253-286.
10. Braun-Blanquet, G. 1951. Pflanzensoziologie. Springer, Wien.
11. Butzer, K.W. & C.L. Hansen. 1968. Desert and River in Nubia. Univ. of Wisconsin Press, Madison.
12. Caron, M. & H.C. 1966. Plantes Medicinales. Le Petit Guide. Hachette.
13. Cloudy-Thompson, J.L. & M.J. Chadwick. 1964. Life in Desert. Foulis, London.
14. Dan, Y., R. Moshe & N. Alperovitch. 1973. The soils of Sede Zin. Israel J. Earth-Sci. 22: 211-227.
15. Dan, J. & Z. Raz. 1970. The Soil Association Map of Israel (scale 1:250,000). Israel Ministry of Agriculture.
16. Danin, A. 1967. A new *Origanum* from Israel, *Origanum ramonense* sp.n. Israel J. Bot. 16: 101-103.
17. Danin, A. 1969. A new *Origanum* from the Isthmic Desert (Sinai), *Origanum isthmicum* sp. n. Israel J. Bot. 18: 191-193.
18. Danin, A. 1970. A Phytosociological-Ecological Study of the Northern Negev of Israel. Ph. D. thesis, The Hebrew University, Jerusalem (in Hebrew with an English summary).
19. Danin, A. 1972. Mediterranean elements in rocks of the Negev and Sinai deserts. Notes Roy. Bot. Gard. Edinb. 31: 437-440.
20. Danin, A. 1972. A sweet exudate of *Hammada:* another source of Manna in Sinai. Economic Botany 26: 373-375.
21. Danin, A. 1974. Notes on the vegetation near Suez and Fayid (Egypt). Israel J. Bot. 23: 226-236.
22. Danin, A. 1975. Living plants as archaeological artifacts. Biblical Arch. Rev. 1(4): 24-25.
23. Danin, A. 1976. Plant species diversity under desert conditions. I. Annual species diversity in the Dead Sea Valley. Oecologia (Berl.) 22:251-259.
24. Danin, A. 1976. Notes on four adventive composites in Israel. Notes Roy. Bot. Gard. Edinb. 34(3): 403-410.
25. Danin, A. 1977. The Vegetation of the Negev (North of Nahal Paran). Sifriat Poalim (in Hebrew).
26. Danin, A. 1978. Plant species diversity and plant succession in a sandy area in the Northern Negev. Flora 167: 409-422.
27. Danin, A. 1978. Species diversity of semishrub xerohalophytic communities in the Judean Desert of Israel. Israel J. Bot. 27: 66-76.
28. Danin, A. 1978. Plant species diversity and ecological districts of the Sinai desert. Vegetatio 36: 83-93.
29. Danin, A. & G. Orshan. 1970. Distribution of indigenous trees in the Northern and Central Negev Highlands. La-Yaaran 20: 115-120.
30. Danin, A., G. Orshan & M. Zohary. 1964. Vegetation of the Neogene sandy areas of the Northern Negev of Israel. Israel J. Bot. 13: 208-233.
31. Danin, A., G. Orshan & M. Zohary. 1975. The vegetation of the Northern Negev and the Judean Desert of Israel. Israel J. Bot. 24: 118-172.
32. Datta, S.S., M. Evenari & Y. Gutterman. 1970. The heteroblasty of *Aegilops ovata* L., Israel J. Bot. 19: 463-483.
33. Eig, A. 1931-32. Les éléments et les groupes phytogéographiques auxiliaires dans la flore palestinienne. I. II. Fedde's Repert. Spec. nov. ref. veget. Beih. 63.
34. Eig, A. 1946. Synopsis of the phytosociological units of Palestine. Palest. J. Bot. (Jerusalem) 3: 183-284.
35. Evenari, M. & Y. Gutterman. 1973. Some notes on *Salvadora persica* L. in Sinai and its use as a toothbrush. Flora 162: 118-125.
36. Evenari, M., A. Kadouri & Y. Gutterman. 1977: Ecophysiological investigations on the amphicarpy of *Emex spinosa* (L.) Campd. Flora Bd. 166: 223-238.
37. Evenari, M., L. Shanan & N. Tadmor. 1959. The ancient desert agriculture of the Southern Negev. VI. Chain-wells in the 'Arava Valley. Qtavim 9: 199-215 (in Hebrew).
38. Evenari, M., L. Shanan & N.H. Tadmor. 1971. The Negev. The challenge of a desert. Harvard University Press, Cambridge, Massachusetts.
39. Evenari, M., E.D. Schulze, L. Kappen, U. Buschbom & O.L. Lange. 1975. Adaptive mechanism in desert plants. In: F.J. Vernberg (ed.) Physiological Adaptation to the Environment, pp. 111-120. Intext Educational publishers, New York.

40. Eyal, M. 1975. Stages in the Magmatic History of the Precmabrian in Sinai and Southern Negev. Ph.D. thesis, the Hebrew University, Jerusalem (in Hebrew with an English summary).

41. Fahn, A. 1963. Dendrochronological studies in the Negev. Israel Explor. J. 13: 291-299.

42. Fahn, A. 1964. Some anatomical adaptations of desert plants. Phytomorphology 14: 93-102.

43. Fahn, A. 1974. Plant Anatomy. Pergamon Press, Oxford.

44. Feinbrun-Dothan, N. 1978. Flora Palaestina. Part 3. The Israel Academy of Sciences and Humanities, Jerusalem.

45. Friedman, J., N. Gunderman & M. Ellis. 1978. Water response of the hygrochastic skeletons of the true rose of Jericho (*Anastatica hierochuntica* L.). Oecologia (Berl.) 32: 289-301.

46. Galil, J. & M. Zeroni. 1967. On the pollination of *Zizyphus spina-christi* (L.) Willd. in Israel. Israel J. Bot. 16: 71-77.

47. Galun, M & I. Reichert. 1960. A study of lichens in the Negev. Bull. Res. Counc. of Israel, 9: 127-148.

48. Ganor, E., R. Markovitz, Y. Kesler & N. Rosenan. 1973. Climate of Sinai. Israel Meteorological Service, Publ. Ser. E 22, 43 pp. (in Hebrew).

49. Ginzburg, C. 1963. Some anatomical features of splitting of desert shrubs. Phytomorphology 13: 92-97.

50. Ginzburg, H. 1964. Ecological anatomy of roots. Ph. D. thesis, The Hebrew University, Jerusalem (in Hebrew with an English summary).

51. Glahn, H. von. 1968. Der Begriff des vegetationstyps im rahmen eines allgemeinen naturwissenschaftliche typenbegriffes. Pflanzensoziologie systematike Den Haag 1-14.

52. Goldberg, P. 1977. Late quaternary stratigraphy of Gebel Maghara. in O. Bar-Yosef and J.L. Phillips (eds.). Prehistoric Investigations in Gebel Maghara, Northern Sinai. pp. 11-31. The Hebrew Univ. Jerusalem, Qedem 7.

53. Greuter, W. 1971. L'apport de l'homme à la flore spontanée de la Crète. Boissiera 19: 329-337.

54. Gruenberg-Fertig, I. 1966. Phytogeographical-Analytical Study in the Flora of Palestine. Ph. D. thesis, The Hebrew University, Jerusalem.

55. Gutterman, Y. 1978. Seed coat permeability as a function of photoperiodical treatments of the mother plants during seed maturation in the desert annual plant: *Trigonella arabica* Del. J. of Arid Environments l: 141-144.

56. Gutterman, Y. & M. Evenari. 1973. Studies of day length on seed coat colour, an index of water permeability of the desert annual *Ononis sicula* Guss. J. Ecol. 60: 713-719.

57. Gutterman, Y. & W. Heydecker. 1973. Studies of the surface of desert plant seeds. I. Effect of day length upon maturation of the seed coat of *Ononis sicula* Guss. Ann. Bot. 37: 1049-1050.

58. Gutterman, Y., A. Witztum & M. Evenari. 1967. Seed dispersal and germination in *Blepharis persica* (Burm.) Kuntze. Israel J. Bot. 16: 213-234.

59. Gutterman, Y., A. Witztum & M. Evenari. 1969. Physiological and morphological differences between populations of *Blepharis persica* (Bornm.) Kunze. Israel J. Bot. 18: 89-95.

60. Haas, H. 1977. Radiocarbon dating of Charcoal and Ostrich egg shells from Mushabi and Lagama sites. In: O. Bar-Yosef and J.L. Phillips (eds.). Prehistoric Investigations in Gebel Maghara Northern Sinai. pp. 261-263. The Hebrew Univ. Jerusalem Qedem 7.

61. Halevy, G. 1971. Autoecology of three *Acacia* species in the Negev and Sinai. M.Sc. thesis. The Hebrew Univ., Jerusalem (in Hebrew).

62. Halevy, G. & G. Orshan. 1972. Ecological studies on *Acacia* species in the Negev and Sinai. I. Distribution of *Acacia raddiana*, *A. tortilis* and *A. gerrardii* ssp. *negevensis* as related to environmental factors. Israel J. Bot. 21: 197-208.

63. Halevy, G. & G. Orshan. 1973. Ecological studies on *Acacia* species in the Negev and Sinai. II. Phenology of *Acacia raddiana*, *A. tortilis* and *A. gerrardii* ssp. *negevensis*. Israel J. Bot. 22: 120-138.

64. Hedge, I. & P. Wendelbo. 1978. Patterns of distribution and endemism in Iran. Notes Roy. Bot. Gard. Edinb. 36: 441-464.

65. Hillel, D. & N.H. Tadmor. 1962. Water regime and vegetation of principal plant habitats in the Central Negev Highlands. Ecology 43: 33-41.

66. Horowitz. A. 1976. Quaternary paleoenvironment of prehistoric settlements in the Avdat/Aqev area. In: A.E. Marks (ed.) The Prehistoric Paleoenvironments in the Central Negev. SMU Press, Dallas 1:57-68.

67. Johnson, M.D. & P.H. Raven. 1970. Natural regulation of plant species diversity. Evol. Biol. 4: 127-162.

68. Johnson, M.P. & D.S. Simberloff. 1974. Environmental determinants of island species number in the British Isles. J. Biogeography 1: 149-154.

69. Kadman-Zehavi, A. 1955. Notes on germination of *Atriplex rosea*. Bull. Res. Counc. Israel 4: 375-378.

70. Kappen, L. & O.L. Lange. 1978. Life forms and physiology of desert lichens. INTECOL 2nd Congress abstracts 2:25.

71. Karschon, R. 1966. Environment and tree growth in the rift valley oasis of Ein Hatzeva. La-Yaaran, suppl. 2.

72. Kassas, M. & M.A. Zahran. 1962. Studies on the ecology of the Red Sea coastal land. I. The district of Gebel Ataqa and El-Galala El Bahariya. Bull. Soc. Géogr. Egypte 35: 129-175.

73. Katznelson, J. 1959. The climate of the Negev. Teva Vaaretz l:310-318 (in Hebrew).

74. Kearney, T.H. & P.H. Peebles. 1960. Arizona Flora. 2nd ed. Univ. California Press, Berkeley.

75. Kislev, M.E. 1972. Pollination Ecology of Desert Plants. Ph.D. thesis. The Hebrew Univ. Jerusalem (in Hebrew with an English summary).

76. Koller, D. 1955. Germination regulating mechanisms in some desert seeds II. *Zygophyllum dumosum* Boiss. Bull. Res. Counc. Israel 4: 381-387.

77. Koller, D. 1957. Germination regulating mechanisms in some desert seeds IV. *Atriplex dimorphostegia* Kar. et Kir. Ecology 38:1-13.

78. Koller, D. 1964. The survival value of germination regulating mechanisms in the field. Herbage Abstr. 34: 1-7.

79. Koller, D. & M. Negbi. 1955. Germination regulating mechanisms in some desert seeds. V. *Colutea istria* Mill. Bull. Res. Counc. Israel 5D: 73-84.

80. Koller, D. & N. Roth. 1964. Studies on the physiological and ecological significance of amphicarpy in *Gymnarrhena micrantha* (Compositae). Amer. J. Bot. 51: 26-35.

81. Koller, D., M. Sachs & M. Negbi. 1964. Germination regulating mechanisms in some desert seeds. VIII. *Artemisia monosperma*. Plant Cell Physiol. 5: 85-100.

82. Laetsch, W.M. 1974. The C_4 Syndrome: A Structural Analysis. Ann. Rev. Plant Physiol. 25: 27-52.

83. Lamprey, H.F., G. Halevy & S. Makacha. 1974. Interactions between *Acacia*, bruchid seed beetles and large herbivores. E. Afr. Wildl. 12: 81-85.

84. Lipkin, Y. 1971. Vegetation of the Southern Negev. Ph.D. thesis, The Hebrew University, Jerusalem, Israel (in Hebrew with an English summary).

85. Litav, M. 1957. The influence of *Tamarix aphylla* on soil composition in the Northern Negev of Israel. Bull. Res. Counc. Israel 6D:38-45.

86. Litav, M. & G. Orshan. 1971. Biological Flora of Israel. I. *Sarcopoterium spinosum* (L.) Sp. Israel J. Bot. 20: 48-64.

87. Lyshede, O.B. 1977. Anatomical features of some stem assimilating desert plants of Israel. Bot. Tidsskrift 71: 225-230.

88. Maarel, E. van der. 1971. Plant species diversity in relation to management. In: E. Duffey & A.S. Watt (eds.). The Scientific management of animal and plant communities for conservation. pp. 45-63. Blackwell, Oxford.

89. Maximov, N.A. 1931. The physiological significance of xeromorphic structure of plants. J. Ecol. 19: 272-282.

90. Mattatia, J. 1976. The Relationships between Amphicarpic Plant and their Environment. Ph.D. thesis, The Hebrew University, Jerusalem (in Hebrew with English summary).

91. McGinnies, W.G., B.J. Goldman & P. Paylore. 1968. Deserts of the World. Univ. Arizona Press, Tucson.

92. Monod, T. 1931. Remarques biologiques sur la Sahara. Rev. Gén. Sci. Pures Appl. 42(21): 609-616.

93. Neev, D. & K.O. Emery. 1967. The Dead Sea, Bull. Geol. Surv. Israel 41: 1-147.

94. Negbi, M. 1968. The status of summer annuals in Palestine. Israel J. Bot. 17: 217-221.

95. Noy-Meir, I. 1973. Desert ecosystems. Ann. Rev. Ecol. Syst. 4: 25-52.

96. Noy-Meir, I. 1971. Multivariate analysis of desert vegetation. II. Qualitative/quantitative partition of heterogeneity. Israel J. Bot. 20: 203-213.

97. Noy-Meir, I., N.H. Tadmor & G. Orshan. 1970. Multivariate analysis of desert vegetation. I. Association analysis at various block sizes. Israel J. Bot. 19: 561-591.

98. Noy-Meir, I., G. Orshan & N.H. Tadmor. 1973. Multivariate analysis of desert vegetation. III. The relation of vegetation units to habitat classes. Israel J. Bot. 22: 239-257.

99. Oppenheimer, H.R. 1960. Adaptation to drought: xerophytism. In UNESCO Arid Zone Research Series vol. 15: 105-138.

100. Orshan, G. 1953. Note on the application of Raunkiaer's system of life forms in arid regions. Palest. J. Bot. (Jerusalem) 6: 120-122.

101. Orshan, G. 1954. Surface reduction and its significance as a hydroecological factor. J. Ecol. 42: 442-444.

102. Orshan, G. 1963. Seasonal dimorphism of desert and Mediterranean chamaephytes and its significance as a factor in their water economy. In Water Relations in Plants. Blackwell Scientific Publ. 206-222.

103. Orshan, G. & M. Zohary. 1963. Vegetation of the sandy deserts in the Western Negev of Israel. Vegetatio 11: 112-120.

104. Otterman, J., Y. Waisel & E. Rosenberg. 1975. Western Negev and Sinai ecosystems. Comparative study of vegetation, albedo and temperatures. Agro-Ecosystems 2: 47-59.

105. Picard, L. 1951. Geomorphology of Israel; Part I, the Negev. Geological survey of Israel.

106. Preston, F.W. 1962. The canonical distribution of commonness and rarity. Ecology 43: 188-215.

107. Quézel, P. 1965. La Végétation du Sahara. Fischer, Stuttgart.

108. Quézel, P. & C. Martinez. 1961. Le dernier interpluvial au Sahara central. Libyca 6-7: 211-227.

109. Raunkiaer, C. 1905. Types biologiques pour la géographie botanique. K. Danske Vid. Salsk. Forhandl. 5.

110. Ravikovitch, S. 1967. Soil map of Israel at scale of 1:250,000, Survey of Israel.

111. Rudich, D. & A. Danin. 1978. The vegetation of the Hezeva area, Israel. Israel J. Bot. 27:160-176.

112. Schoental, R. 1968. Toxicology and carcinogenic action of pyrrolizidinc alkaloids. Cancer Research 28: 2237-2246.

113. Semach, Y. 1975. Ecological studies in *Salsola inermis*. M.Sc. thesis, The Hebrew University, Jerusalem (in Hebrew).

114. Shanan, L. 1975. Rainfall and runoff relationships in small watersheds in the Avdat region of the Negev desert highlands. Ph.D. thesis, The Hebrew Univ. Jerusalem.

115. Shanan, L., N.H. Tadmor & M. Evenari. 1958. Utilization of run-off from small watersheds in the Abde (Avdat) region. Israel J. Agric. Res. 9: 107-129.

116. Shanan, L., M. Evenari & N.H. Tadmor. 1967. Rainfall patterns in the Central Negev desert. Israel Expl. J. 17: 163-184.

117. Sharon, D. 1972. The spottiness of rainfall in a desert area. J. Hydrology 17: 161-175.

118. Sharon, D. 1977. The climate of the Negev. In: E. Sohar (ed.) The Desert, Past, Present, Future. Pp.37-40. Reshafim, Tel-Aviv (in Hebrew).

119. Shmida, A. & G. Orshan. 1977. The recent vegetation of Gebel Maghara. In: O. Bar-Yosef and J.L. Phillips (eds.) Prehistoric Investigations in Gebel Maghara, Northern Sinai, pp.32-36. The Hebrew Univ. Jerusalem, Qedem 7.

120. Smith, A.W. 1972. A gardener's dictionary of plant names. Revised by W.T. Stearn. Cassel, London.

121. Täckholm, V. 1974. Students' Flora of Egypt. ed. 2. Cairo.

122. Tadmor, N.H. & G. Orshan. 1964. Competition between *Avena sterilis* L. and *Stipa tortilis* Desf. under conditions of adequate water supply. Israel J. Bot. 13: 234-245.

123. Tadmor, N.H., G. Orshan and E. Rawitz. 1962. Habitat analysis in the Negev desert of Israel. Bull. Res. Counc. Israel 11D: 148-173.

124. Trease, G.E. 1961. A Textbook of Pharmacognosy, 8th ed. London.

125. Waisel, Y. 1963. Ecotypic differentiation in the flora of Israel III. Anatomical studies of some ecotype pairs. Bull. Res. Counc. Israel 11D: 183-190.

126. Waisel, Y. 1972. Biology of Halophytes. Academic Press, New York.

127. Waisel, Y. & N. Lipschitz. 1968. Dendrochronological studies in Israel. II. *Juniperus phoenicea* of northern and central Sinai. La-Yaaran 18: 63-67.

128. Walsh, G.E. 1974. Mangroves: a review. In R.J. Reimold and W.H. Queen (eds.) Ecology of Halophytes. pp. 51-174. Academic Press, New York.

129. Walter, H. 1964. Die Vegetation der Erde, Vol. I. G. Fischer, Jena.

130. Wendelbo, P. 1961. Studies in Primulaceae. II. An account of *Primula* subgenus *Sphondylia*. Arb. Univ. Bergen. Mat.-Naturv. Serie 11: 5-49.

131. West, R.C. 1956. Mangrove swamps of the Pacific Coast of Colombia. Ann. Assoc. Am. Geogr. 46: 98-121.

132. Wickens, G.E. 1975. Changes in the climate and vegetation of the Sudan since 20,000 B.P. Boissiera 24: 43-65.

133. Winter, K. 1974. Evidence for the significance of crassulacean acid metabolism as an adaptive mechanism to water stress. Plant Sci. Letters 3: 279-281.

134. Winter, K. & J.H. Troughton. 1978. Photosynthetic pathways in plants of coastal and inland habitats of Israel and the Sinai. Flora (Jena) 167: 1-34.

135. Winter, K., J.H. Troughton, M. Evenari, A. Läuchliz & U. Lüttge. 1976. Mineral ion composition and occurrence of CAM-like diurnal malate fluctuations in plants of coastal and desert habitats of Israel and the Sinai. Oecologia (Berl.) 25: 125-143.

136. Yaalon, D. 1963. On the origin and accumulation of salts in groundwater and in soils of Israel. Bull. Res. Counc. Israel 11C: 105-131.

137. Yaalon, D.H. & J. Dan. 1974. Accumulation and distribution of loess-derived deposits in the semi-arid and desert fringe areas of Israel. Z. Geomorph. 29: 91-105.

138. Yair, A. 1974. Sources of runoff and sediment supplied by the slopes of the first order drainage basin in an arid environment (Northern Negev, Israel). Abh. Akad. Wiss. Göttingen Mathematisch-Physikalisch Klasse, III Folge, Nr. 29:403-417.

139. Yair, A. & M. Klein. 1973. The influence of surface properties flow and erosion processes on debris covered slopes in an arid area. Catena 1: 1-18.

140. Zohary, D. 1953. Ecological studies in the vegetation of the Near Eastern Deserts III. Vegetation map of the Central and Southern Negev. Palest. J. Bot. (Jerusalem) 6: 27-36.

141. Zohary, D. 1973. The origin of cultivated cereals and pulses in the Near East. Chromosomes Today 4: 307-320.

142. Zohary, D. & P. Spiegel-Roy. 1975. Beginnings of fruit growing in the Old World. Science 187: 319-327.

143. Zohary, M. 1963. On the geobotanical structure of Iran. Bull. Res. Counc. Israel 11D, Suppl.: 1-113, 1 map.

144. Zohary, M. 1966. Flora Palaestina Part I. The Israel Academy of Sciences and Humanities, Jerusalem.

145. Zohary, M. 1972. Ibid., part II.

146. Zohary, M. 1973. Geobotanical Foundations of the Middle East. Gustav Fischer, Stuttgart.

147. Zohary, M. 1976. A New Analytical Flora of Israel. Am Oved, Tel-Aviv (in Hebrew).

148. Zohary, M. 1978. The Plant World. Am Oved, Tel-Aviv (in Hebrew).

149. Zohary, M. & N. Feinbrun. 1951. Outline of the vegetation of the Northern Negev. Palest. J. Bot. (Jerusalem) 5: 38-53.

150. Zohary, M. & G. Orshansky. (Orshan) 1951. Ecological studies on lithophytes. Palest. J. Bot. (Jerusalem) 5: 119-128.

GLOSSARY

Italics refer to keywords defined eslewhere in the glossary.

ACHENE: A simple, dry, one-seeded indehiscent fruit.

ADVENTITIOUS ROOTS: Roots that arise in aerial plant parts, mostly stems, and not at the pole of young roots.

AMPHICARPIC PLANTS: Plants bearing two types of fruit, differing either in form of ripening time.

ARIDO-ACTIVE PLANTS: Perennial plants prominent throughout the year.

ARIDO-PASSIVE PLANTS: Annual or perennial plants shedding their above-ground parts during the dry season.

BATHA: A *semishrub* formation typical of Mediterranean habitats.

BEDDED LIMESTONE: Sedimentary rock with alternating hard limestone and soft chalky or marly layers.

BIOLOGICAL CRUST: A crust on soil surface composed of fine soil particles and fungi, algae, lichens, mosses or liverworts.

BRACHYBLAST: A short shoot often bearing clusters of leaves; generally with short nodes.

CATCHMENT AREA: The area which directs all the rain water that falls on it, apart from that removed by evaporation, into a wadi which then carries the water to the sea.

CHALK: A soft white or greyish sediment of calcium carbonate; in its purest form it may contain as much as 99 percent of calcium carbonate.

CHAMAEPHYTE: Perennial plant whose *renewal buds* are within 25 cm of the soil surface.

CHERT: A sedimentary rock composed of silica oxide.

CHLORENCHYMA: Chlorophyll-containing tissue in parts of higher plants, as in leaves.

CHOROTYPE: The phytogeographical regions *(phytochoria)* in which the plant occurs.

COLONIZING SPECIES: Plants that rapidly inhabit new habitats.

CONTRACTED VEGETATION: Vegetation which is restricted to wadis receiving substantial quantities of runoff water.

CONVECTIVE RAIN: Rain which is caused by the process of convection in the atmosphere.

CUTICULAR TRANSPIRATION: The loss of water vapor from plant parts through the epidermal cells (see also *transpiration*).

DIASPORE: The seed together with any additional part of the plant which functions in its dispersal.

DISPERSAL UNIT: See diasopre.

DISTAL: located away from the point of attachment.

DISRTICT OF VEGETATION: An area with a typical climate, soil, geomorphology and vegetation.

DISTURBED SOIL: Soil the structure of which was substantially altered largely as a result of human activity.

DOLYCHOBLAST: A long shoot with relatively long nodes.

DROUGHT EVADERS: See *arido-passive plants*.

DROUGHT PERSISTENTS: See *arido-active plants*.

EFFECTIVE RAIN: Rain providing water that can be used by plants.

ENDEMIC: A plant species which is present only in a given region or in part of it.

ENDOLITHIC ALGAE: Algae living within rocks.

EPHEMERAL ROOTS: Roots that function during a short wet period.

GEOPHYTE: A perennial plant having the *renewal buds* under the soil.

GLYCOPHYTE: A plant growing in leached soils (antonym — see *halophyte*).

HALOPHYTE: A plant that grows well on soils having a high salt content.

HEMICRYPTORHYTE: A perinnial plant in which the *renewal bud* is located at the soil surface on top of a storage root or system of thick roots.

HETEROCARPOUS: Producing more than one kind of fruit.

HYDROHALOPHYTES: Halophytes of moist habitats.

HYDROPHYTE: A plant which grows in a moist habitat.

HYGROCHASTIC: Plants that disperse their seeds by rain water.

ISOHYET: A line on a map joining places having equal quantities of rainfall over a certain period.

ISOTHERM: A line on a map joining places having the same average temperature over a certain period.

LANDSAT IMAGERY: Images obtained from information broadcsted from NASA's LANDAST satellite.

LOESS: A deposite of fine *silt* or dust which is generally considered to have been transported to its present position by wind.

MANGROVES: A tropical woodland formation developing in coastal areas and estuaries below the high-tide mark.

MARL: A mixture of clay and calcium carbonate.

MESOPHYTE: A plant requiring moderate amounts of moisture for optimal growth.

MYCORRHYZA: The combination of the hyphae of certain fungi with the roots of vascular plants.

NITROPHYTE: A plant living in soils rich in nitrogen.

PAPPUS: An appendage consisting of a group of hairs, hairy bristles or scales adapted to dispersal by wind or other means.

PHANEROPHYTE: A tree or shrub with *renewal buds* borne high above soil surface.

PHRYGANA: See *batha*.

PHYTOGENIC MOUNDS: Sand accumulated around plants.

PHYTOCHORIA: Phytogeographical region with boundaries corresponding to the distribution of species *endemic* to this region. Four such regions meet in Israel and Sinai.

PIONEER PLANTS: See *colonizing species*.

PLANT SUCCESSION: A gradual process brought about by the change in the plant community's composition during development of vegetation in the area. new species may gradually replace the original inhabitants.

PNEUMATOPHORE: A submerged or erect root that functions in the respiration of *mangrove* plants.

POIKILOHYDRIC PLANTS: The entire plant dehydrates during the dry season and remains in an anabiotic state of desiccation. It resumes activity following wetting.

PSAMMOPHYTE: Plant thriving on sandy soil.

REG: A *silty* soil developing under extreme desert conditions from alluvium and have shallow layers of gypsum, calcium carbonate and salt.

RELICT: A persistent, isolated remnant of a once-abundant species.

RENEWAL BUD: Bud which gives rise to new stems. The location of these buds in relation to soil surface determine the size and structure of the plant.

RESURRECTION PLANTS: See *poikilohydric plants*.

RHIZOSPHERE: The mass of substrate associated with the roots.

SEMISHRUB: See *chamaephyte*.

SILT: Mineral particles of the soil from 0.002 to 0.05 mm in diameter; it is finer than sand but coarser than clay.

STEM ASSIMILANTS: Plants having green stems that carry the main burden of photosynthesis throughout the year; they have diminished or no leaf area as consequence of xeric conditions.

SUBSTRATE: Bedrock or soil supporting plants.

SUTURE: A line of union between closely united parts.

THERMOPHYTE: A plant that thrives at high temperatures.

THEROPHYTE: An annual plant whose *renewal buds* are in the seeds.

TRANSPIRATION: The loss of water vapor by the plant parts, mainly through the stomata.

TWIG: A small branch or slender shoot as of a tree or a shrub.

VESICULAR HAIR: A hair at the top of which there is a small sac filled with fluid.

XENOPHYTES: Plants which arrive from remote regions having no connection with the local flora.

XEROHALOPHYTES: *Halophytes* of dry habitats.

XEROMORPHS: Plants with specific structures adapted to retard *transpiration*.

XEROPHYTES: Plants characteristic of dry and hot environments.

SUBJECT INDEX

PLANT NAME INDEX

Life form, family names and common English name, when available, are given in brackets following the full scientific plant name. Key to the life forms: An — annual; Ch — chamaephyte or shrub; G — geophyte; H — hemicryptophyte; Pg — perennial grass; Sh — shrub; Tr — tree; V — vine. Some desert species may display two life forms, depending on environmental conditions. Family names are abbreviated, e.g. Mimosaceae to Mimos. and Acanthaceae to Acanth. Names of associations in caps., e.g. ANABASETUM ARTICULATAE or *Anabasis articulata - Halogeton alopecuroides* assoc. Whenever a species is described in detail or illustrated the page number is in bold print.

Q

Quercus [Fagac.; oak] 43, 104
Q. *brantii* Lindl. [Tr; Fagac.] 96
Q. *calliprinos* Webb. [Tr; Fagac.; live oak, kermes oak] 83, 104

R

Ranunculus asiaticus L. [G; Ranunculac.; Asiatic crowfoot] 58
Ravenna grass — see *Saccharum*
Reaumuria [Tamaric.] 19
R. *hirtella* Jaub. & Sp. [Ch; Tamaric.] 27, 38, 47, 53, 56, 58, 61, 68, 124
R. *negevensis* Zoh. & Danin [Ch; Tamaric.] 45, 58, 124
REAUMURIETUM NEGEVENSIS 40, 42, 45, 56, 59
REAUMURIETUM HIRTELLAE 37, 40, 42, 47, 56, 59
Reboudia pinnata (Viv.) O.E. Schulz [An; Crucifer.] 124
reedmace — see *Typha*
RESEDETUM MURICATAE 37
Retama raetam (Forssk.) Webb [Sh; Papilion.; white broom] 29, 31, 36, 41, 51, 53, **56**, 62, 68, 69, **93-94**, 127
R. *raetam* — *Achillea fragrantissima* assoc. 40, 56, **57**, 59
R. *raetam* — *Astragalus camelorum* assoc. 50
R. *raetam* — *Rhus tripartita* assoc. 37, 40
R. *raetam* — *Zilla spinosa* assoc. 61
RETAMETUM RAETAMI **55**, 62
Rhamnaceae 119
Rhamnus dispermus Ehrenb. ex Boiss. [Sh; Rhamnac.; two-seeded buckthorn] 43, 68, 72, 104
Rheum palaestinum Feinbr. [H; Polygon.: Palestine rhubarb] 45, **125**
Rhus tripartita (Ucria) Grande [Sh; Anacard.; Syrian sumac] 27, 43, 72, 104, **105-106**, 125
rhubarb — see *Rheum*
Robbairea delileana Milne-Redhead [Ch; Caryophyll.] 41
rocket — see *Diplotaxis*
rocket-salad — see *Eruca*
Rosaceae 82
rose mallow — see *Hibiscus*
rose of Jericho — see *Anastatica*
Rubia tenuifolia D'Urv. [V; Rubiac.; narrow-leaved madder] 58
Rumex cyprius Murb. [An; Polygonac.; rose dock] 124, 126
R. *vesicarius* L. [An; Polygonac.] 124, 126
rush — see *Juncus*
ryegrass — see *Lolium*

S

Saccharum ravennae (L.) Murray [Pg; Gramin.; Ravenna grass] 121
safflower — see *Carthamus*
sage — see *Salvia*
Sageretia brandrethiana Aitch. [Sh; Rhamnac.] 68, 72, 104
Salix L. [Tr; Salicac.; willow] 121
Salsola baryosma (Roem. & Schult.) Dandy [Ch, Sh; Chenopod.; fetid saltwort] 47
S. *cyclophylla* Baker [Ch; Chenopod.] 47, 60, 63, 66, **68**
S. *cyclophylla* — *Cyperus jeminicus* assoc. 66
S. *inermis* Forssk. [An; Chenopod.] 18, 19, 29, 39, 46, **111-112**
S. *jordanicola* Eig [An; Chenopod.] 18, 39, 112
S. *schweinfurthii* Solms-Laub. [Ch; Chenopod.] 58
S. *tetragona* Del. [Ch; Chenopod.] 91
S. *tetrandra* Forssk. [Ch; Chenopod.] 29, 38, 47, 53, 56, 59, 61, **90-91**, 107, 109
S. *tetrandra* — *Hammada negevensis* assoc. 61, 62
S. *vermiculata* L. var. *villosa* Eig [Ch; Chenopod.] 38
S. *vermiculata* — *Satureja thymbifolia* assoc. 37
S. *volkensii* Schweinf. & Aschers. [An; Chenopod.] 18, 39, 46, 112
SALSOLETUM CYCLOPHYLLAE 67
SALSOLETUM RIGIDAE 42, 61, 62
SALSOLETUM SCHWEINFURTHII 56
SALSOLETUM TETRANDRAE 37, 38, 40, 42, 56, 59, 61
SALSOLETUM VILLOSAE 37, 40, 47
saltwart — see *Salsola*
Salvadora persica L. [Tr; Salvadorac.; tooth-brush tree] 18, 38, 39, **50, 63, 64,** 66
Salvia aegyptiaca L. [Ch; Labiat.; Egyptian sage] 38
S. *dominica* L. [Ch; Labiat.; dominica sage] 27, 126
S. *multicaulis* Vahl [Ch; Labiat.] 125
S. *viridis* L. [An; Labiat.; green sage] 32
sand-rat — see *Psammomys*
sand spurrey — see *Spergularia*
Sarcopoterium spinosum (L.) Sp. [Ch; Rosac.; thorny burnet] 27, 37, 41, 81, **82-83**, 119
S. *spinosum* — *Astragalus bethlehemiticus* assoc. 37
S. *spinosum* — *Phlomis brachyodon* assoc. 37
Satureja thymbrifolia Hedge & Feinbr. [Ch; Labiat.] 125
savory see *Micromeria*
saxaul — see *Haloxylon*
Schoenus nigricans L. [Pg; Cyperac.; black galingale] 71
Schouwia thebaica Webb [An, Ch; Crucifer.] 62, 65
Scilla hanburyi Bak. [G; Liliac.; Hanbury's squill] 19
Scolymus maculatus [An; Composit.; golden thistle] 125
Scorzonera judaica Eig [H; Composit.; Judean viper's-grass] 124
S. *papposa* DC. [H; Composit.] 45, **124**
Scrophularia hypericifolia Wydl. [Ch; Scrophulariac.] 62
S. *libanotica* Boiss. [Ch, H; Scrophulariac.; Lebanon figwort] 71
S. *xanthoglossa* Boiss. [Ch; Scrophulariac.] 29
sea-blight — *Suaeda*
sea heath — *Frankenia*
sea lavender — see *Limonium*
Seidlitzia rosmarinus Bge. ex Boiss. [Sh, Ch; Chenopod.] 29, 38, 63, 66
Senniella spongiosa (Muell.) Aell. [An, Ch; Chenopod.] 18, 46
shaggy sparrow-wort — see *Thymelaea*
shrubby horstail — see *Ephedra*
Silybum marianum (L.) Gaerth. [An; Composit.; blessed thistle] 124, 126
Sinai primrose — see *Primula*
Sisymbrium irio L. [An; Crucifer.] 124
Sodom-apple — see *Calotropis*
Solanaceae 126
Solanum eleagnifolium Cav. [An, H; Solanac.] 46
Solenostemma oleifolium (Nect.) Bullock & Bruce [Ch; Asclepiad.] 63
sorghum 54
Spergularia diandra (Guss.) Heldr. & Sart. [An; Caryophyll.; diandrous sand spurrey] 18
Sphenopus divaricatus (Gouan) Reichenb. [An; Gramin.] 121
spurge — see *Euphorbia*
squill — see *Scilla*
Stachys aegyptiaca Pers. [Ch; Labiat.] 72
STACHYDETUM AEGYPTIACAE 56
Sternbergia clusiana (Ker-Gawler) Spreng. [G; Amaryllid.] 16, 41, 45, 58, **100-101**
Stipa capensis Thunb. [An; Gramin.; twisted-awn feather grass, Cape grass] **110**
S. *tortilis* Desf. (synonym) 110
STIPAGROSTIDETUM SCOPARIAE **50, 55,** 62
Stipagrostis ciliata (Desf.) De Winter [Pg; Gramin.] 41
S. *lanata* (Forssk.) De Winter [Pg; Gramin.] 29

INDEX OF GEOGRAPHICAL LOCATION

To locate a place mentioned in the book the reader is refe-rred to the coordinates in Figure 7, e.g., 'Arad is in square G/4 and Dimona in F/4. The English transliteration of place names is based on maps produced by the Survey of Israel.

The meaning of the most common physiographic objects in Hebrew and Arabic are as follows:
Be'er (Hebrew) — well.
'Ein (Arabic) — spring.
'En (Hebrew) — spring.
Gebel (Egyptian Arabic) — mountain; pronounced "Jebel" by the Bedouin of Israel and Sinai.
Har (Hebrew) — mountain.
Makhtesh (Hebrew) — crater.
Nahal (Hebrew) — river or dry water course.
Ras (Arabic) — mountain peak or cape.
Wadi (Arabic) — dry water course.